"Community disaster resilience recently has been utilized frequently and become a relevant framework for scholars and practitioners in the field of disaster management. This book contributes to the growing literature on community resilience with a specific emphasis on resiliency of business community in relation to local governance. The book not only provides theoretical perspectives on resilience but also offer practical insight for professionals in the field."

—Naim Kapucu, *University of Central Florida*

"Understanding business disaster planning is an important topic not only to small, medium, and large companies, but also to local government officials. This text gives prime examples of the impact of planning on the community as a whole."

—Stacey Mann, *Jacksonville State University*

# Toward Resilient Communities

In June 2011, the city of Minot, North Dakota, sustained the greatest flood in its history. Rather than buckling under the immense weight of the flood on a personal and community level, government, civic groups, and citizens began to immediately assess and address the event's impacts. Why did the disaster in Minot lead to government and community resilience, whereas during Hurricane Katrina, the nonresilience of the government and community of New Orleans resulted in widespread devastation?

This book seeks to answer that question by examining how local government institutions affect pre- and post-disaster community and business resilience. Using both survey methods and interviews, Atkinson analyzes the disasters that occurred in New Orleans, Louisiana; Palm Beach County, Florida; and Minot, North Dakota. He argues that institutional culture within local government impacts not only the immediate outcomes experienced during response, but the long-term prognosis of recovery for a community outside the walls of city hall. Understanding tendencies within a community that lead to increased vulnerability of both individuals and businesses can lead to shifts in governmental/community priorities, and potentially to improved resilience in the face of hazard events.

Relevant to scholars of public administration, disaster researchers, and government officials, this book contributes to a growing literature on community and business resilience. It explores not just the devastation of natural disasters, but profiles governmental impacts that led to responsive and able processes in the face of disaster.

**Christopher L. Atkinson** has taught courses in the School of Public Administration at Florida Atlantic University, Boca Raton, Florida. He received his Ph.D. from Florida Atlantic University. His research interests include public management and policy studies, neo-institutionalism, regulation, and emergency management.

# Routledge Research in Public Administration and Public Policy

# Toward Resilient Communities

## Examining the Impacts of Local Governments in Disasters

**Christopher L. Atkinson**

Routledge
Taylor & Francis Group

NEW YORK AND LONDON

First published 2014
by Routledge
711 Third Avenue, New York, NY 10017

and by Routledge
2 Park Square, Milton Park, Abingdon, Oxfordshire OX14 4RN

*Routledge is an imprint of the Taylor and Francis Group,
an informa business*

First issued in paperback 2015

*Library of Congress Cataloging-in-Publication Data*

Atkinson, Christopher L.
  Toward resilient communities : examining the impacts of
local governments in disasters / Christopher L. Atkinson.
      pages cm — (Routledge research in public administration and
public policy ; 8)
  1. Disaster relief—Government policy—United States.   2. Local
government—United States.   3. Disaster relief—Government policy—
United States—Case studies.   I. Title.
  HV555.U6A85  2013
  363.34′80973—dc23
  2013027176

ISBN  978-0-415-65803-4 (hbk)
ISBN  978-1-138-19404-5 (pbk)
ISBN  978-0-203-07630-9 (ebk)

Typeset in Sabon
By Apex CoVantage, LLC

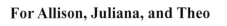

For Allison, Juliana, and Theo

# Contents

# Figures and Tables

## FIGURES

## TABLES

# Preface

The common trait that binds society together now is change, and not always change for the better; risk is all around. The public sector is becoming more aware of the need for risk mitigation. Local governments are under greater scrutiny, and are expected to do more with less. To complicate matters, for a significant portion of the population government continues to be looked at as something of a problem—a leaky bucket that wastes resources and in return gives negligible benefit. Government is exposed to withering criticism from the same public that demands and depends upon its services, creating an untenable situation that is ultimately harmful to government's capacity to respond effectively when called upon. Unfair or not, the government-as-problem mindset typically holds until some terrible event occurs, when government must respond and tie a community, state, or nation together.

The criticism is far-reaching and difficult for the public sector to process. The additional expectations of the public during disaster obscure the clear requirement for public workers to attend to their own needs if they are to be effective in their jobs. Recognition that individual skills, strength, and courage play a role in the quality of institutional responses and in the capacity of government is needed. Public administrators must be accountable, but also responsible—"to profession, one's own conscience, or the public interest,"[1] or all three. To this end, a strong moral compass can guide those who serve the greater good in times of trial; a belief in the role and responsibility of the public sphere, and in limiting or alleviating the suffering of others above self, are central considerations.

Public institutions embody the best and worst of their employees while conveying broader public values. Camilla Stivers suggested that "administrators must use their best judgment together with their sharpest technical skills and deepest sense of what the law demands-the best mind joined with the best heart. This notion is . . . an unenforceable obligation, and one that cannot be solved by diversity or preparedness training."[2] While unenforceable, it is truly an obligation to public service in the best use of that term; the combination of great skill and a sense of responsibility can lead an individual through trauma, stabilize them, and cause them to act responsibly. Public employees are often at their best when society is at its worst.

In a disaster situation, the moral-ethical obligation of the public service takes on critical meaning. Public administration is not a value-neutral proposition. Administration writ large sets out to do the work of the public, but it is not always clear if administrators are merely agents of political forces or are able to set the tone and agenda of government's work. A call to the noble work of public service still exists; bureaucrats should not simply act as paid instruments of the political will, blindly carrying out the desires of the state. This becomes acutely important in disasters and crisis situations, when the public sector is called upon to engage in exceedingly complex operations, under difficult scenarios. Society is dependent on how its institutions and those that run them will function when confronted with the unthinkable. The public's concerns about abilities in this regard, and those of proactive members of the public sector, might settle on applications of resilience, and seek to examine how some communities are seemingly better than others when it comes to crisis management. Such discussions provide us a starting point.

*Resilience* is a term frequently used and abused. It has been an imprecise and amorphous concept applied broadly, to the point of diffusion. However, it could be said that resilience has at its heart the idea that people will go on. As people go on, they grow and reinvent themselves and their communities. We might understand the strength of individuals in crises and how this determines in part their ability to recover, but the connections between individuals and businesses, and public institutions, which are collectives that have tended to be humanized and aped for their failings, remains elusive.

The literature on the subject of resilience has been wide-ranging but lacks in application beyond isolated cases. Theory building and testing is appropriate. While this book reviews resilience as a concept, the focus is first and foremost on the role of government in hazard response and recovery, the interaction of government and its official actors with communities, and how government works with business and individuals to facilitate recovery or fails to do so. This book's project is to seek greater understanding of the connection between our communities, institutions, and ourselves, so that we may have greater control of our fates, and understand more completely why some communities win or lose their fight in responding to hazards.

There is a real opportunity for governments to be at the center of efforts to reduce the potential for losses. Work to interact meaningfully with community partners and provide for a social safety net must occur before, during, and after hazard events, so that the next time a community is threatened, the risk may be lessened. With extreme events on the rise and costing more per incident than ever before, we must begin to address the preventable tragedy of a public that needs desperately the leadership of a strong and moral government with a central devotion to the public interest, which nevertheless struggles with a dislike and even distrust of its public sector.

My hometown, Canton, Illinois, was struck by an F-3 tornado on July 23, 1975; the storm caused two deaths, fifty-nine injuries, and great disruption for 127 businesses.[3] Even today, people in town remember the date and the event, the heat of that afternoon, and the terrible roar of the wind, and they vividly recall the aftermath as if it just happened. The event endures in Canton lore, along with an 1835 tornado that killed the town's founder and his young son.[4] Growing up in a city touched by such events informed my belief that communities must rely upon themselves. Trust and honesty are necessary; people need to know, talk to, and believe in each other, and the time to have such conversations is not when a storm is bearing down or when floodwaters are threatening. Societal changes have led to a decrease in social capital in some places, and this has made communities vulnerable.

Katrina made many reconsider just how prepared they were, and how capable governments were in responding to hazard threats. There was so much anger and hurt in the literature that was coming out of the disaster in subsequent years that there was a need to more objectively understand the nature of the events surrounding Katrina's impact. There is room to expand knowledge in this area beyond simple catharsis.

I am grateful to Routledge's editors and staff and especially Natalja Mortensen and Darcy Bullock, for the opportunity to further pursue my doctoral research program through this book project and for all their assistance. Great thanks to Professor Alka Sapat, who served as my dissertation committee chair at Florida Atlantic University, and to committee members Professor Cliff McCue and Professor Ann-Margaret Esnard for their feedback, advice, and encouragement. FAU provided support through a grant program for research-related expenses for the New Orleans and Palm Beach County case studies. The university provided me a wonderful environment to learn and grow as a scholar and I am indebted to the faculty of its School of Public Administration for their inspiration. I thank all the business owners who took my survey during the original project, as well as the public officials in the New Orleans area, Palm Beach County, Florida, and Minot, North Dakota, for their willingness to take time to tell me more about their experiences running their agencies and how they responded to these events. Thanks to Mr. David Waind, Minot City Manager, who read the draft on the city's flood response and provided helpful feedback.

Special thanks to my wife, Allison, and daughter, Juliana, for their constant patience and cheer; they have been wonderful traveling companions on my research adventures, and I could not have completed this project without them. I also express thanks to my family and friends, and to the following people, in no particular order: Dean David McAleavey, Dean Christopher Sterling, Professor James Maddox, Professor Cynthia McSwain, Professor Jill Kasle, Professor Lori Brainard, Ms. Rosslyn Kleeman, and Mr. Dan Sheterom, who all made an especially positive impact on me while I was at George Washington University. GWU was an outstanding place to learn and the school's faculty inspired me and shaped my view of the world for the better. I also thank Ms. Jackie Kohler, Mr. William Leitze, Mr. Joe Roman, Captain Kenneth R. Force, USMS, Ms. Jean Sefchick, and Ms. Pamela Madison.

Any opinions and views presented here are solely mine and do not represent any institution or organization. With any project of this scope, errors of omission or commission may occur; where they exist, such oversights are mine—please forgive them.

# 1 Local Government Impacts on Resilience in Disaster

In June 2011, the city of Minot, North Dakota, sustained the greatest flood in its history. Rather than buckling under the immense weight of the flood on a personal and community level, government, civic groups, and citizens began to immediately assess and address the event's impacts. The fabric of the community, richly woven of a common experience and sensibility, was and is a driving force behind the response seen in the city, and has been critical in assuring city and regional recovery. The nation may not remember Minot's flood event, as it has largely forgotten Hurricane Wilma and its impacts on Palm Beach County and South Florida—impacts that count the storm fifth on the list of costliest hurricanes. Collectively, we remember Hurricane Katrina, and how a drowned New Orleans spoke to a pressing concern with how governments responded to disasters. While the vulnerability of an entire city left the nation searching for answers, we nevertheless can learn from the event and seek to prevent such disaster elsewhere.

Natural hazards play out on the national and international stage and draw great interest from the public. Coverage of subsequent hazard events, such as various tornado outbreaks, the 2010 Haiti earthquake, the 2011 Tōhoku earthquake and tsunami in Japan, and superstorm Sandy in 2012, has been extensive in media outlets. The coverage may be beneficial in that viewers and readers become more aware of risks that exist, but may also play on emotions and overwhelm, while providing little beyond the stoking of rage through the portrayal of desperate people in dreadful situations. The media tends to stay around a place longer if there is a story; a hazard event, without an accompanying disaster, tends to have a shorter lifespan in media coverage because tragedy makes copy that the mass media can sell. On a personal level, we might think about loss and helplessness, and consider what we would do in those situations; the emotional response to seeing these stories can be visceral. If the media does not delight in having a story to cover, it gains something in dramatic effect for all the human loss.

A difference exists between a hazard event and a disaster. A disaster occurs where the resources and ability of a community, state, or nation to respond are overwhelmed by the scope of an event, a lack of planning on the part of the community, or a lack of available resources to meet the challenges of the event itself. All hazard events do not result in disasters. That disasters provide a great deal more material for the media to cover should be obvious; these events, which exceed our ability to respond effectively, and result in loss, also tend to arouse emotions of anger and betrayal. They speak to a dichotomy within modern society, which desires not only to question government and its methods to the point of vilification, but also to express exasperation when government is not available to protect the public responsibly. If there is a case to be made for government, it is perhaps the role of government when all bets are off, and when a state of nature prevails as it does when disaster confronts a community.

If a clear role for government at all levels exists for responding to hazard events and preventing disasters, it is remarkable that governments are not more certain of their role in response and recovery. Is government's role limited to cleaning up the streets after disaster? Does it extend to helping people and businesses to individually recover? What about contributing to long-term recovery, and allowing a community to regain some sense of normalcy, even if that means a "new normal" that incorporates changes forced upon the community by the event? Sometimes governments are not sure and their responses betray this lack of confidence.

Local government has a responsibility to serve its community by working with state and federal government, community partners, the business community, and individuals to encourage overall community resilience. Municipal and county governments are heavily dependent on the local business community in their bids to recover from disasters, but relatively little attention has been given to business resilience as a distinct element of emergency planning and management. This book includes business resilience as a goal of local government response and recovery, and argues that local governments, through programs and institutional culture, affect business resilience in disasters, which in turn affects community resilience. This has considerable implications for study of disasters—local governments themselves can affect recovery through programs that impact local businesses, as well as through the culture of the governmental institution itself.

## EXPLORING RESILIENCE IN LOCAL GOVERNMENT

While the devastation of Katrina as a disaster event is well-known, we know less about the role of government in affecting community and business resilience. This book seeks to answer the following question: How do local government institutions affect pre- and post-disaster business and community resilience?

In considering this question, we will explore community resilience in disaster from a variety of perspectives. Using both surveys and interviews, we visit New Orleans, Louisiana, Palm Beach County, Florida, and Minot, North Dakota, each impacted by hazard events. Individuals, businesses, and governments work in relationship with and are dependent upon one another, and institutional culture within government impacts not only the immediate outcomes experienced during response, but also the long-term prognosis of recovery for the community outside the walls of city hall. The argument is that local governments matter in impacting community and business resilience.

The effectiveness of a community in addressing and resolving its problems leads to outcomes that may be predictable. Understanding tendencies within a community that lead to increased or decreased vulnerability of individuals and businesses can lead to shifts in governmental/community priorities and potentially to improved resilience in the face of hazard events. This book contributes to a growing literature on community resilience through focus on business resilience, the application of neo-institutional theory, and exploration of the topic beyond individual instances of destruction to profiles in governmental impacts that led to responsive and able processes in the face of disaster.

The goal of resilience is becoming widely recognized as a trait worth pursuing for governments at all levels. However, while governments know that resilience is important, it is not at all clear that governments know how to build resilience in their communities. As with many policy issues, dealing with issues of vulnerability and resilience is prospective, complicated, and vague; this might take secondary importance to more pressing matters of the day. Setting a policy agenda that deals with reducing vulnerability and increasing resilience might be difficult for the general public to understand.

The alternative, which is to wait for a hazard event to, in effect, test the resilience of a particular place, its institutions, and its people, is unacceptable. There are many examples, including Hurricane Andrew and more recently Hurricane Katrina, where poor or nonexistent planning led to increased vulnerability and lack of resilience following the event. Governments benefit from studies of vulnerability and resilience, but isolated case studies illustrating disaster response have been the primary vehicle for research.

Analysis across cases, which explores specific aspects of resilience and vulnerability in a manner that is generalizable, and where the lessons learned can be broadly applied in other communities, holds much promise. This approach can provide immediate benefit to communities that have not dealt with hazard events, as well as increasing the response strength of those that have seen mediocre results. Even when communities are quite different, we might focus on governmental responses, and find that government processes are more alike across cases than they are different. The character and context of place is unique, but the stock activities of governance, such as public procurement or economic development, are not so dissimilar from place to place. Optimal practices matter in how performance is evaluated; understanding this, approaches that have yielded acceptable or strong results provide transferable knowledge.

One of the principle aims of this book is to change the way government looks at its role in reducing community vulnerability and increasing resilience in disaster situations. This rather lofty goal can be accomplished through a street-level look at government, through the eyes of both the community and public officials. A wall between government and community affects how well an area can respond after a disaster, and must be overcome.

## ON GOVERNMENT AND BUSINESS RESILIENCE

Resilience has many definitions. One definition is "a measure of how well people and societies can adapt to a changed reality and capitalize on the new possibilities offered," including ideas of both adaptive capacity and "learning and growth . . . [with] the notion of disaster as a catalyst for development."[1] Speaking of adaptation, people can be resilient in that they respond to the crisis, adjust to the new reality, and succeed, or they can fail to cope, carry the scars of traumatic events with them, and never be able to return to normal. One goal of studying resilience is to understand what makes certain individuals or groups more resilient than others, and to apply this knowledge elsewhere.

In the context of groups such as communities, resilience may be thought of as "a community or region's capability to prepare for, respond to, and recover from significant multi-hazard threats with minimum damage to public safety and health, the economy, and national security."[2] Communities can build resilience by anticipating threats, reducing vulnerabilities that may exist, responding to and recovering from a disaster quickly, and seeking to ensure that institutions do not stand in the way of individual resilience.

Business resilience may be defined as the economic capability of a community's private sector to respond to disaster. Businesses that do not prepare for disasters, of natural or manmade cause, find difficulty recovering from such events; however, businesses do not exist in a vacuum—resilience is community-based.[3] Communities may look upon a disaster event as an opportunity not only to rebuild, but to rebuild differently—"safer, better, and in sometimes more equitable ways" and in a planned approach that improves on what was there before the disaster struck.[4] Balancing these possibilities in outcomes is often a matter for political and economic interests, and some parts of the community may be left underserved through the process.[5] While there is agreement that business resilience is an important concept as an area for study, the literature on business resilience has been limited.[6]

Local governments have a responsibility to communities for business resilience because they are dependent on the success of the business community. Sales taxes and revenue from business operation are tied to the resilience and success of local businesses. If businesses are not successful, whether the failure is the result of the business owner, a disaster, or something else, the government will not derive needed revenue from business activity. It follows then that local governments should work with the business community, before an event, to address concerns and develop a plan for partnerships that reinforce local business resilience.

Government has a coordinating role to play in increasing resilience and reducing vulnerability. It can play this role by fostering openness and accountability within the government itself, providing for fair and competitive procurement, and working with community partners to reduce vulnerability in the business community, especially among small businesses. Public institutions and their cultures have an impact on the business community. They can engage and invigorate business efforts, or fundamentally undermine them. Local government institutions in particular can be a catalyst to help create critical networks that must be in place before a disaster event occurs, that enable community resilience.[7]

Local government's role in disasters is primary and runs from well before the event in planning to the last of the recovery efforts. The importance of local government's choice to support the health and well-being of the local business community could not be clearer. The vision, voiced from the top of institutions, must be carried throughout the structure so that it can inform all aspects of what government does and become inherent in the culture of the institution itself. In this way, business resilience, for start-up firms through large businesses, can be enhanced. Not engaging the issue responsibly will have deleterious effects—a lack of well-paying jobs, little innovation, greater vulnerability, and community weakness and deterioration.

## PUBLIC EXPECTATIONS FOR DISASTER RESPONSE

From a functionalist perspective, disasters "involve demands that exceed capabilities."[8] Ability or inability to manage issues of socioeconomic importance has an impact on the extent to whether an event will be perceived as a disaster by a community. We may expect that government will be called when the demands of the situation exceed individual or business capabilities, but how much are those capabilities informed by a society's comprehension of its own vulnerability? When governmental involvement is requested, will just any government involvement do?

Typically, we think of government in disaster situations in terms of the national government's response. However, disaster response is a multilayered concern that involves authorities from the local government up through the state and federal levels. Local governments are on the front lines where resilience is concerned. Local government responds to disaster first and longest; there is a long-term relationship between community and the city or county itself. Total impact of hazard events is a succession of individual and community-level impacts, and the threat is leveled squarely at communities, businesses, and individuals, perhaps in a more palpable way than if we characterize a threat as national or global.

The threat of disaster is perhaps more certain now than at any point in modern memory. The risks faced from natural and human-induced hazards are ever-present. The year 2011 was the most costly year on record in terms of disaster losses.[9] The following year, 2012, was less costly but only in relative terms, with $160 billion in total losses due to natural catastrophes worldwide.[10] Populations continue to rise, which makes hazard events more likely to impact greater numbers of people. A large segment of the population lives in areas that have substantial natural hazard risks, from coastal regions and fault areas to flood plains. The frequency and intensities of the events themselves make disaster, for lack of a better term, a normal occurrence. Nevertheless, the public remains shocked by it, and many are complacent in the face of the threat.

How people respond to hazards is still a matter of human will and understanding. There is no reason to accept that each hazard event will become a disaster. The tipping point for a disaster is where a response is mismanaged and/or ill-resourced. Professional public administration suggests responses to both of those problems, through strong management rooted in the public good, and proper resourcing of government efforts—food, shelter, and resources for victims of the hazard event. Looking at the tools of public administration alone, responding to a hazard is not really all that different from responding to a budget crisis or a staffing shortage. Detail and rationality of bureaucracy are currency in public administration, and these need not have negative connotation in a hazard context. As much as disasters are normal, it may be possible to normalize our response, and have that response be both effective and responsive to the public. This is a great simplification. However, it does not hurt for us to examine the matter rationally, exclusive of its strained emotional underpinnings, if we hope to shift it back from the realm of emotion and desperation to simply resolving a difficult problem.

The public has high expectations for response to disasters, and government's role in disaster situations has been increasingly called into question. The role of response in a world of crisis is integral—not responding to disaster effectively is a betrayal of public trust. Citizens may believe that governments make people more vulnerable to hazards through a lack of timely and adequate response, or just what appears to be evidence of incompetence in the face of overwhelming odds. A lack of planning, tenuous relationships between community partners, and inundation of a system can look like incompetence. How government reacts impacts how the individuals and groups will perceive and be impacted by a hazard. Communities are made more vulnerable by their own actions, which drive people apart or result in destruction of community, and their failure to recognize and respond to identified needs. There is an economic and resource divide in the world generally. Individuals are themselves vulnerable for a variety of reasons that have little to do with the government itself, like age, race, or gender. As we become more fragmented as a civilization, our response to dire concerns is less focused and more frantic.

When individuals and community groups cannot manage the response to a hazard and recovery of the area, government is called upon to provide resolution. Consider Saundra Schneider's comment on government's role in disaster: "Government is the only institution with the resources and the authority to help citizens cope with . . . cataclysmic events."[11] Looking beyond the role of government response as key, we may also regard government's role as so central that it may turn a hazard event into a disaster, or prevent a hazard from becoming a disaster at all.

We must recall that there is a basis for government intervention in the market in the first place. In disaster, private markets will probably not clean up their own streets, fix the public's infrastructure and dewater all roads, de-muck the public schools, and resolve the community itself into greater employment across industry lines. After all, these are government activities. They relate to public goods and the private market relies upon them as much as individuals. Given the need, government undertakes forward-thinking economic development and disaster planning for community resilience. Government's role is not optional even under normal circumstances: without well-paying jobs, demand in the community's market will not equal supply and the community will not grow.[12]

In a disaster, the role of government institutions becomes even more important. Gentrification may result in the weeks and months following the disaster event, and the resulting community may be changed so much as to become an unrecognizable caricature of its former self. Any community enhancements after disaster must consider fairness to all to be representative;[13] otherwise, rebuilding "the right way" is ultimately one person's political value-judgment,[14] and it may not serve the interests of the community at large.[15]

Governments themselves can be vulnerable. Institutions may have difficulty responding to events that are unusual, great in impact, or where a response is unplanned. Even where events of a certain type are common, it is possible that changes in leadership, or breakdowns in communication within institutions, can prevent a government from responding well to challenges.

## INSTITUTIONS AND INSTITUTIONAL CULTURE AT THE HEART OF A DISASTER

If we think about a government as a group of people acting within a larger community, these government staff people bring to the job of public service their own approaches, beliefs, and backgrounds. The institution sets priorities. The institution itself has a set of assumptions about what it considers to be appropriate and acceptable behavior, and these may not line up well with those evidenced by individuals within the group. The individual is thus constrained by the institution; dissonance is created by actions and behaviors that are counter to expected norms of the group. Another individual might evidence behavior that is much like that of the institution; the institution may encourage this person's behavior and his or her actions. The feelings of the two individuals about the institution would be very different. The institution may allow actions from some individuals, while shutting others down completely, preventing their actions to the extent that they do not align with organizational expectations.

Ideas of appropriateness and institutional constraint and encouragement have been taken up in the literature of neo-institutionalism. Meyer and Rowan, Zucker, March and Olsen, North, Clingermayer and Feiock, and a host of others have written in this area, applying principles of neo-institutionalism to organizations. What we can learn from this is that official actors within organizations are constrained and encouraged by their institutions, and this leads to different outcomes as far as how the official actor responds to his or her role, and how the institution moves forward with its work, or fails to do so. The actors influence the institution and the institution influences the actors. There is a constant reifying and development of the organization through this interaction.

A well-informed government considers the implications of its actions in the context of a disaster scenario, so that it may assist its community in recovery efforts that consider long-term goals, above short-run political and economic returns. This includes the increase of overall economic, and therefore business, resilience. Government institutions, in the neo-institutional framework, can constrain behavior of policy actors within organizations, causing them to act or not act in certain ways. This constraining ability of institutions, especially in the case of disaster scenarios, is not well understood and must be further examined.

## GETTING TO RESILIENCE

Governments are in no position to give in to the threats they face from hazards, whether natural or human-induced. There are too many lives at stake—in view of both mortality and quality of life. Governments can gain a fresh perspective on threats and what types of planning and response best serve the public at large.

A major problem for government is that there are competing priorities. In public procurement for example, the task is complex under normal circumstances. When the system for making purchases is stressed after a hazard event, it may fail completely or work at a suboptimal level. It is difficult to ensure transparency and accountability for buying in a post-hazard event environment. Yet the public expects clarity in processes and wants accountability, and governments should recognize this before a hazard event occurs. There are expectations government will respond proactively, but so much of disaster is unclear that reactive stances might prevail.

If the thought has not occurred to government leaders that a hazard event of some type may impact their communities, the thinking needs to change now. Time and resources spent in planning and engaging the community and partner organizations are investments. This planning has potential to serve not only the community's ability to respond effectively under stress when disaster looms, but also to create a tighter community fabric. These actions change the way individuals see themselves in context of their organizations and while empowering them to make a difference when it matters. Communities that communicate and maintain common threads receive benefits beyond disaster preparedness. Vulnerability is reduced by close community ties and a substantial safety net of people, organizations, and governmental institutions. People are happier when they live in a community that cares about itself and has a common vision of its future.

Government and organizations working properly and coherently to save their communities may not make the front page of the newspaper, but it is clear that this outcome is the most desirable. Consider another option, where the safety net fails for a community's most vulnerable, and the population is prevented from returning to their lives and attaining a sense of normalcy. Where resilience is absent, governments will find themselves hard-pressed to stop the trauma when a disaster event occurs. Full recovery may never be possible.

## PLAN OF THE BOOK

This book situates a discussion of resilience in the context of public administration theory and practice, and applies a conceptual model of community resilience across three cases, to encourage improved understanding of the larger implications of community resilience, and specifically business resilience, and local governance.

Neo-institutional theory has useful application to community vulnerability and resilience. Institutions and their official actors are influenced by the institution—constrained or encouraged by it—and this affects how officials behave. Neo-institutional theory is aided by sensemaking in describing the theory and practice of individuals within institutions and how they behave when confronted with hazards.

It might be thought that institutions matter, generally speaking, but in some cases, the role of institution and institutional culture in shaping the long-term reality of community resilience might not be so obvious. Institutional culture can be a difficult concept to define and measure in practice. Vulnerability as it concerns disaster management is examined, and the competing interests that influence local government outcomes are reviewed. Governments are, in a neo-institutional sense, complex systems of actors and policies, constrained and encouraged by history and institutional norms. This shifting landscape makes predicting outcomes difficult under normal circumstances, but in responding to hazards, the decisions that are being made by complex organizations can lead directly to disasters that extend far beyond government.

Community resilience is seen as a combination of public, private, and individual vulnerability and capacity.[16] In describing business resilience, for example, it is thought that certain businesses, characterized by business ownership, location, and a variety of other factors based on vulnerability models, run a greater risk of adverse impacts after a hazard event. The size of disaster at the level of the individual business, local government institutions, institutional culture within government, and business vulnerability each has an impact on business resilience. Because businesses, individuals, and public sector entities are closely linked, vulnerabilities in one sphere of influence impact other spheres. Likewise, improvements in capacity in one sphere can have positive effects for community capacity in other spheres.

As we explore three local government case studies impacted by hazard events, we analyze across cases the impacts of two public policy touchpoints for local business communities—economic development and public procurement. What are the impacts of these programs? How do they typically work, and how might that effect be different in the context of disaster? Economic development and public procurement are discussed, and the potential impacts of a lack of attention from the local government to areas of resilience, including gentrification, are considered.

Economic development refers to the attraction, retention, and expansion of business in support of growing an economy; the field also includes entrepreneurship programs, financial incentives real estate support to encourage business aims, and technical counseling to enhance competitiveness and growth of businesses. It may be difficult to discern a one-to-one correspondence between programs and their outcomes in employment and increases to the tax rolls. Here, an overview of economic development is provided, focusing on small business programs as one point where governments may actively seek to assist small businesses, through special considerations for contracting as an example.

Of the governmental programs that received attention following Hurricane Katrina, public procurement was one of the few addressed in the academic literature. However, there is room for expanding knowledge by looking more at how public procurement functions, and exploring the concept as both a way for governments to acquire products and services needed, and a mode of expressing accountability and transparency of the government itself. Procurement and its practice may be thought of as a window on the culture of government—what it holds as a priority, the actions it considers reasonable, and how it weighs the interests of stakeholders. The book puts forward a view of public procurement as given to technical rationality, and considers how procurement processes behave under duress in the context of disaster in expanding consideration of "disaster procurement" as a distinct element of the procurement function. Public procurement and corruption are old friends. When times become desperate, decisions are made that may run afoul of procurement's tendency toward increasing professionalization and legitimacy.

As noted in areas impacted by Katrina, local needs may play a secondary consideration when national priorities are considered. Business resilience is directly impacted by the responses of all levels of government after a hazard event, but local government is uniquely placed to advocate for the needs of the local business. Protecting a sense of place in disaster, and specifically in avoiding gentrification and protecting local culture, is contemplated. Without attention being given to the role of local government in responding to disasters, places of significant cultural and historical value run the risk of changing into a homogenized other.

The case studies may represent extreme, typical, and influential cases, as identified in strategies set forward by Seawright and Gerring.[17] The cases of New Orleans, Louisiana; Palm Beach County, Florida; and Minot, North Dakota, were selected because of their potential for within-case learning and cross-case analysis. For the New Orleans and Palm Beach County cases, a section providing context and background for the case leads into a consideration of the business perception of the public sector role in disaster, through a statistical analysis of responses to a survey of business owners that are registered as interested in participating on government contracts. The conceptual model for resilience is employed in each case to identify impacts on resilience. An interview section highlighting qualitative analysis of discussions on disaster response, and the public sector's impact on community vulnerability and resilience, follows. The differences between the perspectives, which speak to the realities, fears, and perceptions of failures on both sides, are striking. For the Minot case, we examine local government responses through a qualitative analysis of primary interviews conducted for this book, in light of existing survey information on economic development and response to the floods among the local business community. The Minot case, involving a smaller unit of analysis relative to the other cases, benefits from learning directly from community leaders.

Finally, we consider the results of the cross-case analysis. We find that the size of the disaster and existing vulnerability impacts community resilience; local government institutional tools can have an impact on resilience; and policy structures and culture of the local government are influenced by the history of the local government and the surrounding community. Individual behavior and community perception are important. Community and institutional culture influence one another, and encourage or discourage resilience in individuals and businesses. The review of cases leads a conclusion which focuses on lessons for community resilience.

Local government considerations can have far-reaching impact, and the prevailing culture of the community is as impactful upon post-event outcomes as it is on the responsiveness of government in times of crisis. Governments are themselves of their communities, and how these institutions respond is wrapped in the essence of the community itself. Vulnerability shows in government as it does in the community. While it is impossible to transfer the intrinsic nature of a community's culture, it is possible for communities and their governments to learn what leads to greater community resilience, emphasize those ideals, and overcome limitations.

Through this approach, we see that resilience is accomplished actively and reinforced constantly. If there is one policy matter that governments may address, and make part of their normal course of doing business, it is including matters of vulnerability and resilience as a priority consideration. The benefits of such an approach are far-reaching. Governments and researchers can change the way communities respond by understanding and addressing community needs in a way that gives an affected population the best chance for success post-hazard. This is not an additional burden. Resilience is the role that government is most equipped to play. The paradox is that government officials and business leaders might not have as much control over outcomes in disaster as they might like to think; understanding the role of institutional culture through the neo-institutional lens can tell us why we lack that control and how we can get it back.

# 2 Resilience, Vulnerability, and Neo-Institutionalism

To provide context for our discussion of community resilience and the role of local government institutions in responding to hazard events and disasters, it is essential to first define and discuss key concepts. A review of resilience and vulnerability and their application to local government in disaster is offered first. Institutionalism and neo-institutionalism are then discussed, situating this work in theory that assists in describing organizational and individual behavior. Sensemaking among official actors, which extends understanding of how culture impacts individuals within institutions in a disaster scenario, rounds out the chapter.

## RESILIENCE IN DISASTER

The definition of the term *resilience* has shifted with its application and is far from settled business. Because of this, we first reflect on a foundational definition in our search for what the term means today. Considering ecological sustainability, Holling wrote that resilience is "a measure of the persistence of systems and of their ability to absorb change and disturbance and still maintain the same relationships between populations or state variables."[1] Later in the same article, the author suggests a need to "keep options open . . . [and to presume] not . . . sufficient knowledge, but the recognition of our ignorance; not the assumption that future events are expected, but that they will be unexpected."[2] This introduces two basic concepts—a system returning to normal or the ability to recover, and adaptive capacity.

Since publication of Holling's article, the term *resilience* has proven to be malleable, perhaps too much so, and prone to imprecision in its use.[3] This is consistent with the findings of Janssen and Ostrom, who "experienced a Tower of Babel in hearing the diverse definitions made [of] . . . resilience, vulnerability, and adaptation."[4] There remains much ambiguity at the connection between these concepts, as Gilberto C. Gallopín identified, in that the terms are related but resilience is not, for example, merely the opposite of vulnerability, and adaptive capacity is not just another way to describe resilience.[5] As Folke notes, research has strayed from the original focus of the Holling work as the concept has been generally applied, neglecting matters of importance such as the interplay between resilience and vulnerability, adaptive capacity, and human-in-environment perspectives.[6]

This is not so much a problem as a way forward, given the exploratory nature of this research, but there is a need for some agreement on definitions, as well as theory building across disciplines. The work of Elinor Ostrom on the sustainability of social-ecological systems, which illustrates how movement in one or two subvariables can shift resilience in a system, is a strong example of work that expands theory. Ostrom warned against oversimplification of complex policy problems when addressing questions of resilience, and the tendency to look for simple answers.[7] While this is sage advice, governments still have a need to act in the present and cannot always await more rigorously tested theory from the academic community. Any research that leads to discussions about how communities can be made more adaptive and resourceful, and less vulnerable, is worthwhile.

A typology of resilience definitions has emerged. The application of the term encourages thinking about how units of analysis—from individuals and organizations, to cities and beyond—respond to surprises and crises. The concept is an inviting idea and cuts across a wide variety of disciplines, from its origins in natural science, ecology, and engineering, to social science and the ability of people and organizations "to cope with external stresses and disturbances as a result of social, political, and environmental change."[8]

Southwick and Charney's comprehensive definition of individual resilience, which centers on being "bent, by traumatic experiences . . . but not broken," recognizes that resilience is not merely the absence of vulnerability, but also the presence of an adaptive nature. Resilience is "complex, multidimensional and dynamic."[9] A spiritual ability for weathering trying emotional circumstances, while being able to think creatively and skillfully, may aid resilience.[10] Obstacles abound that confound translation of this dynamic from the personal level to the governmental or institutional level. For example, Sapat observed that "major obstacles can surface in generating the political will necessary to adopt policies to develop resilience, especially for projects that are costly in the short-term and whose benefits manifest only over an extended period."[11] Elran commented that "until we measure societal resilience and collect and compare the relevant data . . . we cannot know which types of communities are most resilient."[12]

Researchers have had difficulty agreeing on what the resilience means, other than that it is important and deserves attention. Small contributions at theoretical margins might confuse, rather than encourage, policy development and planning, and this has possibly limited the contribution of resilience to policy. Rather than reducing uncertainty, vague descriptions of the concept increase confusion while reducing the potential that local governments will engage in serious discussions of the matter. With the lack of consensus in definition, time required to develop resilience in a community, and political will needed to increase it, how is government to encourage its development?

Resilience is a difficult sell to communities because this vague concept must compete for dollars, resources, and interest against other activities. Community leaders understand intuitively what researchers study—that while the relationships between vulnerability, adaptive capacity, and resilience are important, ambiguity is unsettling to elected leaders and the public. As much as it is difficult to explain for researchers, the task is more challenging for public officials, who arguably have more of a vested interest in turning the concept into policy action to save lives. Further, programs that lack champions and do not directly connect dollars to an immediate and concrete increase in welfare are often portrayed as wasteful. The nature of resilience is that it is a local issue; different locales have unique approaches to resilience which might create positive effects through decidedly different paths. This complicates comparisons of one community's resilience to that of another even when ample information is available. Regardless, conversations about social processes and inherent capacity to recover are highly salient and worthwhile. Even normative discussions alone are beneficial, because of the integral nature of the sensemaking component in organizations, discussed later in this chapter, and how individual-level actions interact with institutional culture.

Elran's approach to resilience in communities provides us a starting point: "a resilient community would flexibly contain a traumatic experience, would expeditiously adapt to its consequences, and would bounce forward to an improved functional level."[13] This work evaluates resilience by examining three layers: the first being the incidence of traumatic episodes, the behavior of the public being second, and the public's perception being the final layer.[14] This builds upon other great work in the handling of crisis situations, including Fink's still-relevant foundational text, which stresses the potential for crisis as opportunity, the importance of honest, open, clear communication, the value of leadership, and the need for proactive responses.[15]

Using the Elran model, incidence of events should be a factor in understanding community resilience. Some communities see more frequent hazard events and have attained a level of skill in responding to their own particular crises that allows stability. The behavior of the public is important—a public that is more knowledgeable about potential risks that are posed by hazards should respond more effectively when forced to react. The collective need to go on and recover is not equally felt in all communities. In some places, seemingly the only choice is to pick up and move forward—people are so completely invested in themselves, their neighbors, a sense of self, and a work ethic, that rebuilding quickly is the only legitimate option that presents itself. As contagious as this positive motivation appears to be in some places, discontent is also contagious;[16] where citizens are unhappy, dissatisfied, and not moving forward, soon all may be that way. Public perception, or "personal condition, such as mood, anxiety, and confidence;

perceptions of personal, community, and national states of security; feelings about routine conduct; the rate of optimism concerning one's individual and societal future; fear of terrorism/other hazards; attitude of solidarity and willingness to continue living in the community/country; trust in the local/military/national leadership; confidence in one's personal/community preparedness; and pride in the nation and country,"[17] represents the final layer of evaluating resilience.

Our domination of our environment extends as far as nature will allow it. As societies grow wealthier, to borrow a turn from Browning, their reach exceeds their grasp.[18] They build further with their risks subsidized, and are not able to withstand the force of nature when it inevitably comes into conflict with humankind's growth. When disaster calls, citizens look to government, and are troubled when government's responses are, put kindly, highly variable in effectiveness.[19] Wealth alone does not make society more resilient. The relationship among wealth, hazard, and the potential for loss has been explored and research challenges such assumptions.[20]

Disasters are more than hazard impacts and the destruction of property, and more than disruption of normal schedules for a few days. The impacts can be felt in an interminable displacement that contributes to a sense of helplessness. This is true for both individuals and businesses. To this end, disaster manifests itself in the breakdown of the social fabric, or what Gilbert refers to as the "relaxation of social and political boundaries."[21] Horlick-Jones offers that disasters are socially constructed and evidence themselves in a "loss of social control," and suggests that the media has played a role in enhancing myths about public behavior during disaster situations, which in turn has affected how the public anticipates and responds to such events.[22]

Government clearly plays a role in assuring that the social fabric of a community is preserved throughout the aftermath of a storm. When government cannot or will not fulfill that role, the result is disaster in the fullest sense. "Disaster threatens and involves destruction or disintegration of the extensive, orderly patterns that bind together the large space and many places [of] modern material life. Disasters are problems that are, by implication and in fact, out-of-control, in that they break out of the modern mould, or challenge its effectiveness."[23]

Complicating matters for government is that not every locale has the disaster management skill set it needs to respond effectively to the challenges it may face. Disaster response and potential for recovery hinge on variables in play in the community before the storm hits: whether the community sees the warning signs of risk, the atypical nature of the event itself, and whether the community can learn from prior mistakes.

Paton and Johnston indicated that "knowledge of hazards; shared community values; established social infrastructure; positive social and economic trends; partnerships; and resources and skills" have an impact on community resilience.[24] Regarding the role of trust in the community, they wrote that "trust in the municipal and government authorities and in the private sector is critical [for resilience]; there is a need for high levels of trust and social capital within networks and across networks with an active program to restrict conflict and to generate mutual respect and understanding."[25] They also recommended that smaller community networks be connected with other networks, to avoid isolating individual groups, and that communities seek to keep lines of communication open for all stakeholders.[26]

Alesch, Arendt, and Holly explored the factors affecting community recovery in the aftermath of disaster, and correlated the following factors highly with economic ruin after such an event:

1. *Massive damage and massive systemic effects;*
2. *A weak pre-event economy;*
3. *Non-local owners;*
4. *Inadequate quality or quantity of workers;*
5. *Inadequate infrastructure;*
6. *Increased costs of doing business;*
7. *Heavy penalties for rebuilding (having to meet stringent environmental standards in newly built facilities, for example); and*
8. *A deteriorated location relative to other places.*[27]

The authors found that the groupings of businesses in the aftermath of storms are typically small businesses that serve the community and larger industries that make up the economic engine of the community—tourism, for example, or manufacturing.[28] Small businesses are generally affected in their survival by limitations in their ability to recover assets; their access to suppliers, employees, and customers (the business must have all three to survive); and adaptability to post-disaster circumstances.[29]

Governments frequently make planning missteps in the aftermath of a hazard event, and this has potential impacts for well-being in communities from an economic perspective. Governments generally all want to rebuild and recover, but they do so through approaches having little to do with recovery as a concept of its own: "creating industrial parks, advertising for clean industry, providing all the tax breaks permitted by law, and endorsing chamber of commerce efforts" are all standard economic development tools and may be considered business-as-usual, rather than the sort of post-event toolkit from which governments might most advantageously draw.[30] Put another way, the influx of funds into an area in the post-event environment should not be considered an opportunity for local governments and private sector groups to put together wish lists and argue for a self-serving agenda.

When looking specifically at governmental action, the focus has varied widely. One important study in 1985 looked at fourteen disaster events and found three criteria that the leadership of local governments should have to lead their communities back to recovery from a disaster: *personal leadership, the ability to act,* and *knowledge of what to do.*[31] Others suggest that strong community ties; strong links between government, community institutions, and resource providers outside the community; and public participation make all the difference in a strong recovery.[32] Perhaps not surprisingly, much of the focus in the early part of this century was on the impact of terrorism, rather than natural disaster, on people and businesses. We are perhaps more likely to see a text on the economic impacts of terrorism than on the economic impacts of natural disaster.[33]

There are two types of resilience: inherent and adaptive. Inherent resiliency comes from not exhibiting traits associated with business vulnerability. The adaptive form comes from some businesses' ability to cope and innovate in the face of a threat to survival. Instead of giving up, some businesses become highly creative, change their approach to their business, explore new options of products or services to sell, and in general do what is necessary to continue on in a post-disaster community.[34] It is vital to understand how this will work in practice in disaster—a government has its best opportunity to assist the public and encourage resilience by reducing vulnerability through strategic planning with a broad array of community partners, before a hazard event occurs. This speaks directly to inherent resilience. If a community waits until a hazard event occurs to engage in this dialogue, individuals and their businesses will be forced to fend for themselves. While the human race is remarkable in its ability to adapt, the outcomes from such an approach will not serve the community as would an approach based on increasing inherent resilience by reducing overall vulnerability. Community fracturing can occur when the local government fails to engage such issues.

Community fracturing adversely impacts potential for resiliency. An illustrative example from the literature is Miami-Dade County after Hurricane Andrew. Miami, it has been said, "was ripe for disaster when Andrew struck, both because its vastly increased population lacked recent experience of major storms, and because the community had suffered serious economic reverses combined with the fragmentation of its political structure into fractious groups that lacked a perception of common interests."[35] "After hurricane Andrew moved away, its impacts became grist for the pre-existing interest group coalitions in Greater Miami, which sought to respond to the new realities that it had unleashed."[36] The interest groups that formed as a result of the hurricane included coalitions that evidenced the pre-event community, as well as the new reality of the post-disaster reality; groups formed included race and ethnic coalitions, place-based coalitions, and production-based coalitions.[37] Where inherent resilience lacked, the public took to adaptive resilience to survive.

## THE ROLE OF VULNERABILITY

"Proponents of vulnerability as a conceptual explanation take the position that while hazards may be natural, disasters are generally not."[38] Communities and citizens, and by extension their businesses, frequently do not plan for disasters;[39] people and institutions of all types are brought back to the basics of survival by such events. These are jarring experiences, where nearly every aspect of one's reality may change. Vulnerability is a central issue in any review of community resilience—a concept that ties the community that was, before the event, to the community that may be rebuilt. A move away from the primary role of government has left open the question of security after disaster, and has led to problems in maintaining community support networks, as Watkins offers:

> *Neoliberalism extols the power of markets to solve social and economic problems . . . The neoliberal approach to disaster policy reflects a more general approach to government. Minimize government involvement; where possible, privatize . . . The Katrina disaster is marked by a collapse in social capital. Katrina indicates that "social networks and the associated norms of reciprocity" either did not exist, or were insufficient. Even if networks exist, a lack of material goods may render social capital ineffective.*[40]

The myth of effective private-sector provision of public services undermines growth and maintenance of public institutions that can actually provide effective services and do so steadily and accountably. The analysis to judge benefit for cost of privatization is too often not as thorough as it should be.[41] The theoretical case for such approaches may be "weak or nonexistent."[42] Proponents of market forces point to a wider failure of government to meet public needs, even as they admit that the private sector does not have all the answers. The argument of which is worse—inconsistent government intervention or faulty market responses—is silly and irresponsible. Markets fail as do governments. Reliance on either/or has left communities in peril, susceptible to vulnerabilities in all their forms.

"Vulnerability can be defined as the likelihood that an individual or group will be exposed to and adversely affected by a hazard."[43] It is possible to define differences between vulnerable places and vulnerable populations, though they rely upon similar factors: "biophysical characteristics (e.g. earthquake frequency, climatic variation) and societal characteristics (e.g. social inequity, access to education, poverty)."[44] Approaching vulnerability involves a consideration of social processes and interactions with the environment; vulnerability occurs at the intersection of demographics, reality as it is constructed in communities, and the impacts of hazard events on individuals and groups. Vulnerability can be adjusted and outcomes altered—by changing vulnerability, it is possible to ameliorate adverse impacts before an event even occurs. "Aftermaths of disasters often facilitate the adoption of new hazards policies and plans by individuals and/or coalitions with vested interests. Such initiatives can, in turn, change the vulnerability of society to future events."[45]

"In addition to a hazard, there must be some vulnerability to the natural phenomenon for an event to constitute a natural disaster."[46] Vulnerability places the emphasis on the human role. Vulnerability relies "on what renders communities unsafe, a condition that depends primarily upon a society's social order and the relative position of advantage or disadvantage that a particular group occupies within it." Certain populations live in a marginality "that makes . . . life a permanent emergency . . . a set of variables such as class, gender, age, ethnicity, and disability that affects people's entitlement and empowerment, or their command over basic necessities and rights."[47] Being exposed to a hazard is only part of the equation.

One of the hallmarks of vulnerability usually called to mind is wealth or lack thereof, and the idea of rich and poor in a world that many feel is becoming more and more unbalanced in this respect. The market acting as society's method of striking a balance—creating the greatest good for the greatest number through most effective and efficient use of resources— arguably has resulted in there being no balance. In disaster situations, this is even more apparent; wealthier citizens "allocate their assets more efficiently in dealing with their risks, and manage their risks with minimal welfare loss."[48] Stratification extends even beyond a rich-versus-poor model, to "racial, ethnic, political power, and gender lines," with disaster "exacerbating preexisting inequality."[49]

Vulnerability among small businesses is closely tied to the vulnerability of small business owners. Small businesses frequently lack cash flow, and are not able to fully capitalize their ventures. Financing for the business is sometimes caught up in the personal finances of business owners. More recently established businesses may lack capacity to overcome disruptions. The well-being of a firm and its longevity may have much to do with the owners' "class, gender, age, ethnicity, and disability," to borrow from Bankoff's list of social indicators that ultimately contribute to vulnerability and susceptibility of businesses to a variety of considerations in the face of disaster.[50] Disaster theory holds important implications for business resilience, but additional research is clearly needed.

Vulnerability among businesses, as with individuals, has to do with whether the subjects have obtained a working network of resources. Without a network, businesses and individuals are at considerable risk. It would follow, then, and be worth further examination, the idea that government efforts to support small businesses in the wake of disaster will work only if the community has developed networks.[51] The effect of community might shield businesses from the damage and disruption of a natural hazard's impact.[52]

Vulnerability involves both place and social factors. The place where a business is located can make the business vulnerable. It is not merely that businesses in California must deal with the earthquake threat or businesses along the Gulf Coast and the Eastern seaboard must contend with hurricanes; even the building from which one operates can be a source of vulnerability.[53] Business decisions, such as whether one rents or owns property, the competitiveness of one's specialty field, the necessity or discretionary nature of the product or service being sold, and the presence or absence of community standards, such as building codes, all contribute to business vulnerability.[54]

The characteristics of business owners are important to consider; whether a firm is minority- or women-owned is important because research has shown that those firms are more vulnerable after disasters because they are already more vulnerable to "shifting economic trends."[55] Some research has suggested that African-American business owners make less money from their businesses than do other minorities and Caucasian business owners.[56] When coupling these concerns with the fact that small businesses are most vulnerable to changes in the economy under normal circumstances, it is clear that small minority- and women-owned businesses deserve special consideration. Hurricane Katrina, given the impact on New Orleans and its large African-American population, cast some light on this subject, but to date not enough research has been done on business resilience there or elsewhere.

It has been noted in the literature that "female business owners may have more trouble acquiring loans than their male counterparts."[57] In the case of Hurricane Andrew in Miami, "disaster so starkly exposed both the vulnerability of women and the traditional male power elite that women's groups across the political spectrum were able to identify common issues. Business organizations saw the needs of female small-business owners ignored and worked with religious women leaders active in tent cities serving Miami's poorest women."[58]

While wealth alone does not explain outcomes for resilience, there can be no doubt it is a contributing factor. Again using the example of Hurricane Andrew, Dash, Peacock, and Morrow found that "housing, job, business, and tax revenue losses were proportionately greater in the minority community. At the same time, the poorer community was less able than its more affluent counterpart to manage recovery efforts in the post-disaster period because of major personnel and organizational problems."[59] "Normal disadvantage in political and economic structures was further crippled by . . . lack of experienced administrators and staff as it attempted to deal with complex problems of recovery."[60] Worldwide, higher affluence is also shown to be a key to lowering vulnerability to disaster.[61] Financial inequality leads to differences in vulnerability.[62] These findings tend to support the notion of disparate impacts due to vulnerability and an accompanying effect on resiliency in disaster management. Marginal populations, in terms of day-to-day living before the storm, may have little chance of returning to a pre-storm level of capacity and quality of life in the aftermath.

Local governments are seen as having a first obligation to "disaster recovery assistance to businesses," and with small businesses, the need to reopen quickly is great not only for the business, but also for local governments, who miss these firms and their tax payments.[63] For individuals, addressing functional and access needs in advance of a hazard event will improve the local government's ability to serve those populations, and communicate with them in a way that is clear and makes sense.[64]

## INSTITUTIONALISM AND NEO-INSTITUTIONALISM

In this section, institutional and neo-institutionalism are examined as providing a framework for examining organizational behavior; the influence of institutions and institutional actors upon one another is also addressed. The relationship between sensemaking and neo-institutionalism provides a crucial additional layer of detail to analysis of organizational behavior, which is helpful in understanding decision-making during crises.

Institutionalism is "a general approach to the study of political institutions, a set of theoretical ideas and hypotheses concerning the relations between institutional characteristics and political agency, performance, and change . . . [institutions] are collections of structures, rules, and standard operating procedures that have a partly autonomous role in political life."[65] Institutionalism in its original form relied on comparative approaches of formal structures and was heavily descriptive. The importance of strong and balanced institutions has roots in Hobbes, Locke, and Montesquieu, among others.[66] Organizations are mindfully brought into being by people to meet a specific need, and run by people to meet that need.[67] The institution itself is the focus of the analysis, rather than the people, who while serving some important task in coordinating the function of the larger organization are interchangeable, as the organization is less driven by or given to the whims of their discretion. Institutionalism focused on how government was influenced by the law, the importance of structure and how it influenced behavior; the approach utilized comparisons of whole systems one to another, and strongly rooted its normative analysis in history.[68]

Some of the theoretical assumptions of the tradition are not unassailable. That "actors have stable preferences and thus evaluate the outcomes of individual choices according to stable criteria,"[69] for one assumption, does not fit well with reality, as both preferences and choice criteria are subject to change over time. Also, that the decisions that should be made to optimize outcomes are somehow obvious to the actors involved is a shaky position to hold in practice. The rational choice approach that pervades basic economic analysis does not handle well the concept of motivation, nor does it comprehend the environment and optimize decisions based upon it. It does not consider the more complex questions that arise, that do not involve decisions we make repeatedly and do not have to think about, or more impersonal interactions, such as those that occur in surprise or crisis situations. We fumble for answers, rely upon ideology, and lack important information to make optimal decisions.[70]

"The goal is not to find quick solutions driven by individual choices. Rather, it is the creation of shared interests and shared responsibility."[71]

Value-rich notions such as this, central to normative impressions of public service, for example, are absent in the rational, cold approach suggested by behavioralism. There is nothing about the creation of an institution that implies a solution will be efficient—only that it might provide some sense of stability and thus reduce uncertainty; pretensions to omniscient knowledge of efficiency are, at best, misplaced.

As for why new institutions arise in the first place, the tendency is clearly away from "design" and toward evolution: "the assertion that man has created his civilization and that he therefore can also change its institutions as he pleases . . . would be justified only if man had deliberately created civilization in full understanding of what he was doing or if he at least dearly knew how it was being maintained. The whole conception of man already endowed with a mind capable of conceiving civilization setting out to create it is fundamentally false."[72]

Behavioralism and rational choice demanded that institutionalism respond more effectively to produce theory beyond this basic description of institutions. The countering view went that "any theory of the behaviour of bureaus that does not incorporate the personal preferences of bureaucrats . . . will be relevant only in the most rigidly authoritarian environments."[73] Scholars such as Selznick pushed recognition of the role that influences outside an organization can play upon internal processes. William White extended this to show that actors within organizations are pulled in a variety of ways by considerations within the organization and outside it.[74] Activities of the public sector do not always serve the interest of the public, and institutions sometimes do not constrain the behavior of actors to implement policy as might be most advantageous under a given set of circumstances.

Government had been characterized by supposed "omniscience, benevolence, and limited omnipotence" in institutionalism.[75] As thought on the subject of the role and study of government progressed, institutionalism in its original forms was found wanting; its focus on forms of institutions had largely left out the role of institutional actors. Institutions are, of course, not actually omniscient, nor are they particularly benevolent or omnipotent. Institutionalism did not account for the political opportunism and governmental failures that were being seen, while rationalism and other behavioralist approaches tended to consider these factors more.[76] Institutionalism alone does not explain the actions of collective institutions or the activities of actors within; theory had failed to capture organizations influencing and influenced by their environments. What was needed was realism; what good was theory if it bore no resemblance to the experience of either institutions or official actors?

"Organizations, despite their apparent preoccupation with facts, numbers, objectivity, concreteness, and accountability, are in fact saturated with subjectivity, abstraction, guesses, making do, invention, and arbitrariness . . . just like the rest of us."[77] Where classical institutionalism is focused more on the role of the institution, neo-institutionalism, also called new institutionalism, takes into greater consideration the role of the actors within the system. Neo-institutionalism looks at organizations in their value-laden contexts, adding substance from more behavioralist models of thinking but without rejecting old institutionalism outright. Neo-institutionalism suggests that "actors pursue their interests by making choices within constraints,"[78] and does not assume humankind as "something standing outside nature and possessed of knowledge and reasoning capacity independent of experience."[79] Instead, neo-institutionalism provides an understanding of the teleological, purposive, and behavioralist, even where bounded rationality does not necessarily allow for conscious maximization of benefit to self. It could be said that neo-institutionalism is informed by, and builds upon, the idea of bounded rationality, while emphasizing the interactive role played by institutions and actors.

While coverage of the growth and evolution of neo-institutionalism is beyond the scope of this book, several salient points further our discussion. The foundation of neo-institutionalism is generally thought to be around 1977, with papers by Meyer and Rowan, and Zucker.[80] First, as Greenwood, Oliver, Sahlin, and Suddaby note in their comprehensive introduction to the subject, neo-institutionalism tends to hinge on the importance of organizations at least appearing to behave rationally, and conforming to what Meyer and Rowan referred to as the institutional context, "the rules, norms, and ideologies of the wider society."[81] Second, organizations have competing priorities which do not always line up well, so this conforming behavior may sometimes be symbolic or limited;[82] organizations have to address pressures from different groups and as a result may act in one manner and talk in another,[83] or mediate pressures by creating organizational identities.[84] Third, how institutions behave ultimately has a great deal to do with the concept of legitimacy, including its acquisition and utilization.[85] Finally, neo-institutionalism calls researchers to see actors within organizations as influencing their organizations, as well as being influenced by them; this interaction of institutions and institutional actors has given rise to useful questions about the role that professions and professionalization play in organizations.[86] We will touch on all of these issues at various points later; first, we engage in a discussion of neo-institutionalism at a root level.

There are three basic aspects to neo-institutionalism:

a) Actors pursue a broad set of self-interests, but with limited knowledge and cognitive capacity.
b) Institutions are defined as the rules, combined with their enforcement mechanisms, that constrain the choices of actors. These rules include the laws of states, the policies of organizations, and the norms of social groups.
c) Institutions ideally constrain actors such that their best choices are consistent with the collective good, enabling, for example, mutually profitable exchange between actors.[87]

Neo-institutionalism seeks to find out the "norms of institutions as a means of understanding how they function and how they determine, or at least shape, individual behavior."[88] What is it about institutions (policies, systems of regulations, programs) that constrains behavior and encourages policy actors to act a certain way—to behave *appropriately*? Or if these systems are absent, how do the policy actors behave, and what are the consequences for vulnerability and resilience in a community? Such discussion moves considerably beyond "value-maximizing" theories and onto a two-way street with the institutions that are built in society. Even though self-interest, rather than benevolence, informs many human actions,[89] society nevertheless creates institutions that are not entirely given to personal ends.

Within neo-institutionalism, March and Olsen define an institution as "a relatively enduring collection of rules and organized practices, embedded in structures of meaning and resources that are relatively invariant in the face of turnover of individuals and relatively resilient to the idiosyncratic preferences and expectations of individuals and changing external circumstances . . . Institutions empower and constrain actors differently and make them more or less capable of acting according to prescriptive rules of appropriateness."[90]

In one sense, "institutions derive from the optimizing decisions of individuals and respond to changes in the set of relative prices that individuals face."[91] "Together with the standard constraints of economic theory," institutions "determine the opportunities in a society."[92] Further, "both what organizations come into existence and how they evolve are fundamentally influenced by the institutional framework. In turn they influence how the institutional framework evolves."[93] Institutions represent "the rules of the game in a society, or, more formally, are the humanly devised constraints that shape human interaction . . . institutional change shapes the way societies evolve through time and hence is the key to understanding historical change."[94]

The structure may predispose an institution to producing certain results or otherwise limit what actors can accomplish; some institutional structures are favored by policy actors because they are seen as helping the actor to accomplish his or her intended ends.[95] These rules "shape and are shaped by human behavior" and provide "incentives for political exchange."[96] However, this assessment and the resulting outcomes are imperfect and variable in practice.

As rule consequentialism would have it, a rule system that contains constraints is the "most beneficial" type of system. The thinking is, though a rule system may allow favoring of individual goals, and that pursuit of those goals toward personal ends may outweigh value maximization for the whole, a public system that had no constraints would lead to "consternation" in the system.[97] Public institutions, particularly in disaster situations, are expected to act appropriately and to abide by institutional constraints, as a result, to maximize value for the most people possible. Society has expectations, and these form an institutional context. An institution gains credibility and resources if it meets society's expectations or at least appears competent; resources are withheld from agencies that do not conform to such expectations.

Within the institutional context, the ability of institutions to shape and constrain behavior is an essential aspect of neo-institutionalism. "Institutional constraints . . . are the framework within which human interaction takes place . . . perfectly analogous to the rules of the game in a competitive team sport. The rules and informal codes are sometimes violated and punishment is enacted . . . some teams are successful as a consequence of constantly violating rules."[98] Rules may be taken as "fact" and implemented, or ignored for purposes of implementation because the institution, at an informal or even formal level, designates that as an appropriate response. The idea of appropriateness asks us to consider what individuals would do under normal and new circumstances, or even how we respond in a crisis, and how what is appropriate may conflict with personal preferences.[99]

The need to explain a supposed homogenization of processes and organizational types was initially a vital pursuit in neo-institutionalism,[100] but further examination began to show the diversity of approaches in process and form that exist in practice. Given variations in process, form, and culture, among other organizational traits, essentially the same institutional structure or procedures can be imposed in two unique places, and because of the importance of both the formal and informal institutional processes that are at work, and how they constrain or fail to constrain the behavior of individual actors within the system, the outcomes can be entirely different.

Normative neo-institutionalism focuses on how the norms of institutions shape behavior and inform how an institution runs.[101] Historical institutionalism holds that what occurs early on in the organization's history impacts consequent decisions, and suggests that it is very difficult for an organization to affect marked institutional change of any significance.[102] The common thread among all forms of neo-institutionalism is simply that institutions—the formal and informal structures, institutional actors, and the interaction with forces exogenous and endogenous—matter, and have direct bearing on policy outcomes.

A possibility exists that gradual changing of agencies over time dismantles the state in piecemeal fashion, or causes society to lose some potentially positive attributes, such as the less self-serving attitude that may be seen in some disaster situations. For example, treating the public more as customers than citizens may ultimately result in citizens feeling that they have no responsibility to other citizens, when in actuality citizens owe each other quite a lot in a civil society—at least if they wish it to work. The reminder is a good one—that "to be a citizen requires and commitment and a responsibility beyond the self; to be a customer requires no such commitment and a responsibility only to oneself."[103] Citizens can and should assist each other when that becomes necessary; the duties and responsibilities of a citizen ought not to be optional, to be picked up only when personally beneficial.

The role of institutional actors within neo-institutionalism illustrates the complexity of the policy landscape. Actors in official situations may cooperate or find themselves in conflict with one another; each player is, to greater or lesser extent, weighing the costs and benefits of various actions leading to a dynamic landscape of both policy and practice. Where it is in the interests of all players to cooperate for the benefit of the group, there will likely be cooperation leading to positive, productive outcomes and increased efficiency. It is possible that actors will act in a certain way because it is the purported *right thing to do*, but even what constitutes the right response in any given scenario may be relative depending on institutional context and expectations of the community surrounding the institution, or even the larger society. What is appropriate in the institutional context might be wrong when more broadly considered.

Actors are socialized by their institution, not only according to the identity of the organization but also their role within it, and increasingly understand what is appropriate and what will lead to consequences; this informs their individual behavior[104] and can lead to the institution gaining or losing legitimacy. The institution's legitimacy depends on how well it accomplishes its goals, but also whether it can conform to laws and social norms and values from the institutional context.[105] Legitimacy might also have to do with whether the agency's work and actions are taken for granted by the institution and its context and largely accepted, though it should also be noted that it is probably not necessary for an agency's activities to be considered legitimate by a broad segment of society for the agency to be successful in achieving its goals.[106]

Neo-institutionalism has also been applied to the idea of professionalization. Leicht and Fennell write that "peer-oriented professional practice is under pressure from institutional constituents interested in lower costs, more accountability, and ethical transparency at the same historical moment that technological changes put pressure on traditional institutionalized methods for delivering professional services."[107] As discussed in chapter 4, professionalization and legitimacy in public procurement and economic development are at once influenced by, and influencing, the operational context of government, in a scenario that corresponds closely to neo-institutionalism. Further, pressures from the institutional context have forced both specialties to redouble efforts to show how the knowledge and expertise of these fields adds value in an increasingly technological environment characterized by tight budgets and political pressures.

The literature suggests a relationship between sensemaking and neo-institutionalism.[108] Sensemaking involves making sense of what occurs in life and giving context to events; essentially people generate a social world and then figure out what it means, iteratively.[109] People create meaning through the mental constraints in their structured environments. Sensemaking is central in discussions about disasters because when institutional actors are good sensemakers—that is, they can figure out confused or deteriorating situations rapidly—outcomes can be better for the public. Where identities from organizations prevent such adaptable sensemaking, forcing actors into approaches that insist on developing responses from a limited repertoire of possible options, disasters can result.[110]

"Institutions function to contextualize sensemaking by imposing cognitive constraints on the actors who do the sensemaking."[111] When we think about sensemaking in neo-institutionalism, in a disaster situation, all are seeking to make sense of the situation given incomplete, inadequate, or even wrong information, and constantly adjusting their approach to the situation as they understand it more. It matters little whether the discussion centers on first responders or managers of a local government enterprise in the wake of a hazard event; what matters is the ability of the actors involved to promptly make sense of the situation, adapt, marshal resources, and deploy their system of response as quickly and as meaningfully as possible.

Institutions function over and around the sensemaking abilities of individuals. In some ways, neo-institutionalism continues to suffer some of the same conceptual limitations that affect institutionalism, and sensemaking highlights these limitations. What seems like an organization as a whole acting a certain way, implying a behavior of the collective, is more properly the behavior of individuals engaged in sensemaking of a difficult scenario. These actors take in information, understand some of it and lack needed information to make the ideal call, but they must make a decision at some point. The decision might be wrong, or at least not the most advantageous under the circumstances. Outside observers might be tempted to aggregate the individual behaviors of all concerned and call that an institutional response, but how accurate or detailed is this observation?

In applying theory, decisions may be attributed to the organization that made them, but in reality, key decisions are made under duress by a few individuals, who through a sensemaking process sought to make the best decisions possible under the circumstances, whatever best meant to them at the time. The decisions are attributed to the institution and all actors within. This echoes ideas of neo-institutionalism—that the actors are also influencing each other and being influenced not only by the institution but also the environment around them, which includes an atypical hazard event. The operating context may include agencies and knowledge of how matters have been handled historically, political wants and needs, past failures or successes, and cultural tendencies, whether personal to the sensemaker or unique to the institution, to the extent that individuals might embody the traditions of the whole. The public makes little differentiation between the institution and the individual actors within when accountability is demanded for the outcome. The institution takes the critical hit for individual failures and vice versa.

Weick warns that managers "forget to think in circles,"[112] and suggests that most decision-making processes involve causal paths that are not one-way, but circular. Planning affects disaster response, but disaster response also affects planning. Institutions socialize institutional actors, but institutional actors reinforce socialization within the institution. This is why an organization can be effective at decision-making, but still have poor results; the actors within cannot make sense of shifts in their situations quickly or effectively enough to create a reality within which they may work, even with a top-down strategy. Prior experiences do not guide the actors—and they may believe that the organization itself will lead them through the trauma when it will not. *Appropriateness* may deter resourcefulness. The institution is waiting for the actors to make sense of it all, and if they cannot, disaster can result. Experience of a disaster is not an indication that a future event will be handled more effectively; a disastrous outcome is a warning of the potential for even greater disaster and more serious consequences should additional events occur.

Institutional actors must be able to improvise; wisdom comes from interacting, within and beyond one's immediate circle, and benefitting from a sense of trust and honesty among actors. In more fully developed situations where the learning cycle is complete, it is less likely that the actors within the organization will fail because they are able to adapt and improvise, and because they trust each other.[113] Trust, "a general belief in the good will of others and their commitment to participate in dialogues for negotiating new realities," is essential because everyone in a crisis is vulnerable to some extent.[114] People need to reduce their uncertainty to function well; ambiguity occasions sensemaking and increases the need for trust among official actors.[115]

In sum, resilience and vulnerability are intricate ideas that necessitate new and adaptable ways of thinking about how governments respond to hazard events. Theory has potential to assist in the review and analysis of institutions and capacity to respond in complex scenarios, but only to the extent that theory can adequately reflect the intricacy of public systems subject to myriad influences from within and outside organizations. Neo-institutionalism is a way of looking not only at the "rules of the game," as they are set out in policies and regulations, but also in how the institutions—the formal and informal structures of the organization that implements the policies or regulations—impact the outcome of an institution's work through constraint or encouragement of individual behavior. Institutions have a role to play in setting rules and norms, and how they inspire or inhibit the stability of an entity is a theoretical basis with wide application. The application of such theory has strong potential for use in discussing how communities confront hazards.

# 3   Local Government Institutions in Disaster
## Context and Complexity

Every disaster is local. The mechanisms of government and interaction between local, state, and federal officials, or how nonprofits can work together to solve the problems of response and recovery, are probably not the first thoughts that come to mind when thinking about disaster impacts. Disasters are intensely personal affronts that devastate homes, cities, schools, and families. Poorly managed hazard events leave an open wound that deserts an angry and fearful public. The public may not know at whom or what it should direct its anger, and that only seems to make matters worse.

This sense of public mistrust intrudes on government in practical terms and interferes with both the professionalism of the public service and the ability of public sector employees to perform under unforgiving circumstances. It is unsophisticated to believe that institutions, by virtue of being called governmental, are imbued with a sense of will and expertise that transcends the people within. Governments as institutions are, in the deontological ideal of public service, composed of people obligated to serve all citizens and best consequences. Government in action is the knowledge and skill of those that fill public positions, hopefully representing the best and brightest expert-practitioners government can attract. Government responses to criticism have tended toward a defensive posture, which is evidence of the gulf between responsible citizenship as instrumental in the function of effective government, and pretensions to customer service, which may undermine the role of government in a way that is more evident in times of crisis. Some activities of government are confronted suspiciously by the public. Where government needs to act, it is often hesitant because of the possible repercussions. Criticism of public efforts to protect communities can backfire when government services are needed desperately. Government does not help itself when it withholds needed information or fails to explain well its rationale.

In this chapter, the role of government is considered with respect to public expectations and outcomes, as well as internal and external influences and constraints. The discussion begins with the complex operational environment of local government. We move on to a review of businesses, government, and the public in the milieu of a hazard event or disaster; the interaction of competing interests and its negative impact on planning and preparation is examined. Gentrification as a potential outcome of redevelopment after disaster is viewed as a result of competing interests that fail to consider wider community impacts. Finally, a model for business resilience in disaster which draws upon theoretical bases, and considers the role that local government might potentially play in reducing potential for great harm to communities, is presented.

## PUBLIC EXPECTATIONS AND THE LOCAL GOVERNMENT CONTEXT

Government service asks for selfless individuals who place public service in especially high regard. As an ethical consideration, personal responsibility and the goodness of human nature are topics that have been argued for centuries, and are an undercurrent of public service philosophy. Political forces recognize that expertise is needed to run government, but the activities of public sector experts, ideally acting in an independent and nonpartisan capacity, do not always yield outcomes acceptable to the political sphere. As much as government needs to provide and enforce the rules, so that all interests can play fairly and optimize their outcomes on their own terms by individual success or failure, government can sometimes not perform this role in a way that maximizes the public good. Politicians might be seen as supporting the bureaucracy only as much as needed to accomplish their own personal ends. Political forces cut funding and blame bureaucracy for failures caused by a lack of resources. There can be viciousness to the interaction. This is unfortunate because the dysfunction leads to resentment of government, of government intervention in protecting the public, and of the use of policy tools and planning to reduce impact and harm from hazard events. One might be a bit naive in defending government as it is when it could be argued that the system naturally provides less than optimal outcomes. The public makes little differentiation in resenting politicians and administrators, and sees the public sector as a whole falling short.

Local government works in this challenging environment and has an added responsibility that comes with proximity to events. Local governments are concrete representations of the public sector in our lives. Elementary and secondary education, garbage collection, and building permitting are all manifestations of local government that are tangible and directly experienced. The public is most familiar with these interactions, as they touch community life on a daily basis.

In times of disaster, local government is the first into the destruction and last out. Local government's response is central; with state and federal responses primarily being invoked upon the failure of local government to handle a problem, local governments are expected to be on the front lines of disaster response. However, the quality of response by local government is often a matter of resources and planning, and too often local governments find themselves underprepared to address the significant community needs that accompany hazard events. Dollars, political will, and planning expertise necessary to respond fully might be in limited supply. Too many unknowns and variables exist for any city or county to be fully prepared. A legitimate, responsive, and transparent local government in normal operating mode can seem incompetent and foolish in the face of disaster, because disasters require far too much of government systems in too short a time.

The national government and the states collectively set the direction for the nation as a whole. However, local government resources are first brought to bear on the response and recovery. The concept is that local government should know best what each city/county needs, and have planned for the event in question enough to cover the cost.

This is often an absurd assumption, considering that local governments have copious and financially burdensome requirements placed upon them by the national and state levels. Many local governments lack deep pockets and savings of recovery funds to the extent necessary to make themselves whole again after a hazard event. There can be little expectation that most communities are in a position to fund a response and recovery effort beyond a low-level disruption.

At the local level, problems in responding to crisis are most obvious and excuses used at other levels of government ring hollow. Local governments do not have the luxury of claiming that they were looking out for the public interest of a whole nation and were therefore unable to give the situation the full time and resources it needed, as would the federal government. Both federal and state levels frequently blame local government for a lack of initiative. However, mandates from the state and federal levels clearly cascade down to cities and counties, moving them toward increasingly unsustainable budgets dependent on the whim of returns from property tax rolls. Decisions made at other levels of government impact every city and town in determining what government will value and what resources it will have available to respond to the unknown. While political needs are significant, the ramifications of such policy choices must be considered when they increase vulnerability at the local level.

Local government administration may be seen merely as a means to various ends, with politicians, interested third parties, and enterprising administrators all seeking to influence the priorities and agendas of governance in a manner that leads to haphazard and sub-optimal results. To the extent that this jumbled dynamic leads to inferior ends, efforts should be made to adjust and focus on a longer view in protecting communities. Placed in trying circumstances as a matter of course, local governments are interested in maximizing the use of their resources and getting the greatest utility for expenditures. The utility in providing for the safety of a community through planning efforts might be questioned in a period of limited resources. Self-aware local government must address more formally the ends it wishes to achieve for its population, in increasing quality of life and reducing vulnerability. This may take the form of comprehensive planning and disaster exercises to address hazards; preferably, this planning should take place well before any disaster event, and involve a broad cross-section of the community, especially including the business community.

## DISASTERS, GOVERNMENT, AND BUSINESS

Much of the discussion that occurs in disaster research has to do with what a disaster is, and when it actually occurs. A renunciation of responsibility for order—and more specifically for government's role in maintaining the undergirding of a system of supportive networks that make up a community—may lead to dire consequences. "The first, official concern of the authorities and the larger role of the military and police services in most disasters is to restore order."[1] This is the overarching concern. As much as nature can stress public systems and private interests, people can make disasters worse. Because "individuals ignore a problem because they do not think it will happen to them and believe the public sector will rescue them if it does, there are grounds for imposing requirements in advance of a possible catastrophe."[2]

It has been suggested that community resilience is the resilience of individuals and businesses; the relationship between business recovery and individual displacement has been noted.[3] Both government and business contribute to order in a society. As a result, government and business should have consideration for their societal context and interest in protecting it, day-to-day and especially in the context of a disaster. Neither exists in a vacuum and there is a need for government and business to work together to create stronger communities. "Consideration of the role of the private sector in natural hazard mitigation acknowledges that while business shares responsibility with government for the protection of lives and property, the priorities of business are different from those of government. While government is expected to establish policy for the health, welfare, and safety of the people, business as an economic entity acts to satisfy the demands of the marketplace as a higher priority."[4]

However, employment and the marketplace itself are subject to disruption when health, safety, and general welfare are threatened; business cannot operate in a manner that acts outside society, given that its means and ends are linked to communities. Understanding the differences and the potential for complementarity is essential. Business has a responsibility to act in a manner that first protects its own bottom line and strategic interests. There are some activities, such as moving large volumes of goods over great distances and distributing products, that large business performs in a manner superior to government. Government is perhaps also not one's first choice for feeding people in a disaster area, but business has great capability in that respect if its reopening is included among initial response priorities. Large businesses have varying degrees of disruptions on a daily basis, and have developed methods of being more responsive when subject to those stresses; government can learn from business and how it responds, in many respects. Business in turn can learn from government, and contribute to the greater good for no other reason than to be a part of the community upon which it depends.[5]

There should be greater comprehension of how involved the business community is in disaster response and recovery, among individuals and governments. Businesses more likely will first protect their own interests; this amounts to a concern for the marketplace and is not wholly objectionable. From the perspective of public choice, we might reasonably expect as much. However, businesses under stress may engage in moves that protect their own interests at the expense of the larger market, causing further problems. Businesses may also fail to coordinate with each other and discuss common concerns in a crisis situation.

Government can protect its bottom line—protecting the public—by learning businesslike adaptability and resourcefulness, including bringing business partners to the table in planning for and carrying out disaster response and recovery. Government also has a role in encouraging responsible risk-aware behaviors on the part of small businesses, pre-disaster, given their unique vulnerabilities and the closeness of personal outcomes with the fortunes of these entities; the special risks that small business owners face in responding to hazards are complex and difficult to convey effectively, given an uneven level of business acumen. Businesses and individuals should recognize that government has a role in society and is essential.

Protection of the marketplace and position within it is vital, but such attention to business concerns is only sensible if a marketplace remains. This is equally important to businesses, government, and individuals. Business, without government in such a case, is unable to proceed, because essential order has not been maintained. Business may need the protection of government to reopen and regain stability in the days and weeks following a hazard event. A business cannot open its doors without employees, or may not be able access means of production if supply routes are disrupted.

Business is dependent on government to establish order—to insist upon the rules for which all will be held to account, to allow for a functioning society. This invokes a commitment between government and business. The suggestion that business is not dependent on functioning government is an arrogant position that is hard to defend; the proof lies in how little business is done when government fails to hold up its end of the bargain and enforce the rules. Where business has sought to remove involvement of government in protecting the public interest, the price to pay later has sometimes been far greater than allowing for government's role all along. This is surely the case in environmental protection and the fouling of the commons, and in protection of consumers. What benefits one firm may cause great harm to other societal components and put the whole at risk. Government shields the marketplace, environment, and employment base upon which the business community relies. The business community should embrace being involved in such efforts, as community resilience is not the limited province of government alone.

Community resilience can be considered a public good, one that the market may fail to provide in the normal course of doing business. As a public good, resilience is expected to be provided by government. Its consumption is nonrival—everyone benefits from a resilient community.[6] It is also subject to free-ridership—a business could claim that it derives no value from residing in a resilient community that has planned for disaster, and refuse to pay anything for such benefit, even though it clearly derives benefit through the readiness of its government and employees. The result is that insufficient resources may be allocated to protection from disasters. This is one reason why communities fail to plan for disaster and protect themselves; it is not somebody else's problem to prioritize and pay for such planning, it is everybody's responsibility. Not only is resilience a public good—it could be argued that it is a merit good. Quite for the reason Musgrave states, "the apparent willingness of the public to provide for a second car and a third icebox prior to assuring adequate education for their children,"[7] society must concern itself with providing for resilience because people cannot be depended upon to do what is right in prioritizing it.

A number of nations, the United States among them, since the Enlightenment have had a foundational element in the invisible hand theory of Adam Smith, and notably the pursuit of self-interest above some idea of public interest:

> *Every individual necessarily labours to render the annual revenue of the society as great as he can. He generally, indeed, neither intends to promote the public interest, nor knows how much he is promoting it. . .By pursuing his own interest he frequently promotes that of the society more effectually than when he really intends to promote it. I have never known much good done by those who affected to trade for the public good.*[8]

While the public interest and private interests do not completely align, the public interest may still be served by the private sector acting in its own interests. Society relies upon the private sector to be the engine of prosperity—and nothing that government can do will ever balance a lack of success in the private sector for communities in need of jobs. However, the concept of "divergences between social and private net product,"[9] called externalities, plays havoc with the invisible hand theory. In pursuit of private gains, we can see significant, if not unacceptable, societal costs, which are not often taken up by the corporations that cause them. Our collective well-being may be impacted by the private pursuit of gains for individuals; from an economics perspective, private interests that do not pay the full, socially responsible costs of production are paying far less than the actual costs of their participation in a market. The gap between market cost and actual cost is funded by the government, to pay for cleaning up environmental degradation, public health, or consumer protection, among other spillovers.

Let us extend the Pigouvian dilemma of externalities to the idea of disasters. If we take the example of a city in a risk area for hazard events, which does not plan for disasters or pay for the costs associated with this planning, including improved infrastructure to withstand risks, the costs of operating government might provide for a steady state of operation absent the challenges that would be posed by hazard events. Private interests pay taxes and fund this level of service, but the level of service will be insufficient should a hazard event befall the city. The costs to return to normal will be borne by government, if they are addressed at all, because the revenue stream to fund the city's operations was inadequate to the actual costs of running the city. The government cannot afford these additional costs, and will look to other levels of government for assistance. People will want to return to normal and will demand that their institutions, including the private sector, do whatever is necessary to allow that return to normal. If recovery funds are diverted or misused, communities will never return to normal and vulnerability to future events will increase.

Corporations rely upon public infrastructure; beyond this, they rely on a functioning city, region, state, and nation, for customers, employees, access to distribution channels, and other services. The protection and maintenance of public infrastructure and functional communities is not cheap. It is up to political forces to communicate and insist upon the necessity of resilience to the community.

Because these full costs consider broader issues of public interest, there may be little will on the part of the public sector to engage private interests in discussions about fairness and societal responsibility in preparation for hazard events. Considering the real costs of an environment for business and the public, which include risks, potential hazards, and the need for planning and infrastructure, may not be a popular discussion for politicians and community leaders, but it is a necessary conversation nonetheless. The costs of government failing to fill societal gaps are too large to bear.

"Choices related to levees, floodplains and wetlands, trees and plants, construction standards and roads, health programs and evacuation plans, complex financial instruments, and other matters increase or decrease the potential costs of catastrophes."[10] Government makes those choices, hopefully in partnership with citizens, business, and interested groups, or those choices are made for us through omission; these decisions require political will and the ability to stand up to interests with short-sighted and self-serving goals. The benefit to the broader society in making a choice that reduces or alleviates risk is a shared merit good. Human involvement through politics, including agenda setting for short-run private sector gain, to avoid making difficult decisions to protect a community, or through shoddy administration of public programs, shows vividly the nature of the problem. Making sustainable, responsible, and resilient community choices can be costly and requires coordination, strategy, and political-administrative muscle. This is not an exception to use of political skill and capital—it is a reason for it. It is challenging for a sense of public interest to overcome human nature and a tendency toward entropy in the face of ever-expanding private interests; nevertheless society is not tied to catastrophe as its only outcome, if political forces can utilize skill and expertise to set an agenda that protects the community as a whole.

Giving up is not an acceptable alternative. Barak Orbach raises the specter of Katrina and points out, "the government may fail, including in adequately preparing for or responding to catastrophes . . . the alternative to government action may still be worse."[11] Not only must government be involved in disaster preparation and response—it must be involved in ways that are more consequential, create a heightened state of awareness of true risks, and bring communities together in preparation and partnership through planning, whether or not the majority of the public recognizes the need. Even then, the role of institutions must be understood more clearly.

Institutions are not all of one mind—the vast array of moving parts provides for both opportunity and danger in everyday situations, including regular hazards that might evolve with disastrous consequences. The importance of the informal organization is clear—how employees think and respond within organizations is key. Personnel do not function entirely by rules, but rather by personal relationships, between peers, management, and the public, and are affected by "needs, attitudes, motives" in the workplace.[12] Motives might be as simple as an employee not wanting to be terminated, and knowing that a certain interpretation of an administrative rule will anger those in power—a resulting decision is perhaps not the decision the employee would otherwise make, if left to his or her own devices. As noted in the previous chapter, an institution that lacks strong sensemaking ability may lack adaptability and yield outcomes that fail marginally or massively, depending on the context. It could be said that informal organizational culture influences individual sensemaking and encourages or discourages resourcefulness.

The idea of the technological (or man-made) disaster is common in literature on disasters as well as industrial failures.[13] A pathological culture, for instance, "suppresses warnings and minority opinions," while bureaucratic culture sees people "not encouraged to participate in improvement efforts."[14] An article describing the commonality between the loss of the space shuttle Columbia and Hurricane Katrina noted that being too focused on solving problems can lead to a failure to prevent them in the first place: "identifying problems and solving them, not focusing on how the problem developed."[15] Because sensemaking is iterative, unless the organization encourages thoroughness in thinking through problem prevention, preventing problems before they happen may not be intuitive enough to be properly addresses. Communities have a variety of problems that could be prevented, if only institutional actors were able to make sense of what those problems might be in enough time—to anticipate them and keep them from happening.

The preferred informal culture, which allows organizations to avoid disasters, is "generative . . . able to make use of information, observations or ideas wherever they exist in the system, without regard to the location or status of the person or group having such information, observation or ideas. Whistleblowers and other messengers are trained, encouraged, and rewarded."[16] The free exchange of ideas, even if it challenges existing norms, is encouraged and rewarded. A reliable culture of ideal type offers "autonomy for workers; a questioning, skeptical attitude; an emphasis upon safety; professionalism; and skills."[17]

However, organizations frequently do not know how to find what has gone wrong in the first place. In reality, catastrophes are rare and the rareness of catastrophes leads people to complacency.[18] Culture is probably not the strongest indicator of catastrophic potential in an organization, but it is a contributing factor. It is impossible to plan for all random events, so organizations should not plan to avoid them, but to endure them.[19] Regardless, some organizations are culturally more prone to disasters than others; the antithesis, that certain organizations are safer or less prone to disaster than others, is also true.[20] Informal structures within organizations can have dire consequences, themselves contributing to disaster.[21] For example, a lack of transparency can enable corruption,[22] while groupthink can lead to policy fiascoes that organizations should work to avoid.[23]

Culture may have a bearing on the outcomes seen in the business community after disaster. Institutional culture is a product of the environment in which the institution operates. The institution itself is of its community. The community, in its interactions with its government, creates or at least makes possible the informal culture that exists within the institution among its actors.[24] There is extensive evidence that both formal and informal processes within organizations can have far-reaching implications in outcomes. It is reasonable to conclude that the resilience of the local business community may be impacted by the institutional culture of local government.

Local government institutions, in their approach to the business community, define the potential for resilience and quickness of recovery in the aftermath of disaster events. They do this primarily by understanding the connections between resilience and vulnerability, and reducing vulnerabilities to foster development that is adaptive and capable of success even under distressing circumstances. In a disaster situation, the role of institutions becomes more complex, though give-and-take between actors and institution remains. Institutional actions or the inability to act can make a great difference in outcomes for a community.

In examining the intersection of disasters, government, and business, we have explored the foundational elements of influence upon government acting in a hazard response situation. The environment of government is complicated, and can be made more complex by difficult circumstances. Theory can assist in clarifying how actors and institutions respond under duress; neo-institutionalism is useful in approaching disaster research, because institutional actors do not always act rationally or in ways that can be explained in solely economic terms. Other considerations that are central to the lives of institutional actors at all levels of organizations, including "cultural values, political demands, and aspects of social recognition . . . legitimacy,"[25] take precedence at various times and constrain or encourage behavior. One strength of neo-institutionalism is that it allows a much more thorough portrayal of the complexity of the public administrator at work, and this additional detail in portrayal—where actors are not merely interested in the cost or marginal benefit in monetary terms of taking this action over that—may lead to greater accuracy and improved understanding.

## THE INTERACTION OF LOCAL GOVERNMENT AND BUSINESS—PUBLIC INTEREST OR COMPETING INTERESTS?

There are undoubtedly situations where the market cannot and will not respond quickly enough in any kind of organized or strategic way to be able to provide for a response that would serve the needs of the broad citizenry. Government's role may be to intervene in situations where the market cannot respond to a situation, as in disasters—where there is a clear market failure in the traditional sense. The ability for government to plan and coordinate is essential.

Private entities acting in the public interest is an area of research that continues to be addressed by researchers in the business and society (B&S) literature. This area of study seeks out "causal links between social responsibility and profitability."[26] The concern of whether businesses can act in the public interest, and then if so, whether they do act in the public interest and why, is beyond the scope of this volume. However, we may take from this branch of study that business has had an interest in achieving a level of compatibility with the broader aims and expectations of civilization. It is worth mentioning, primarily because the popular sentiment may suggest otherwise, that social interests of protecting society are not mutually exclusive from a company having a profit motive.

The most important thing that a business can do for its community is reopen, provide jobs, and help to return the community to normalcy, as quickly as makes sense given the circumstances. There are assuredly cases of companies that look beyond their walls to serve the community; for example, Hancock Bank was praised as an example of community-focused action after Katrina.[27] But even Hancock's staff recognized that doing the right thing could result in great profit. This is an example of how the invisible hand worked well in a disaster situation, but it does not work that well in every case, nor would a few positive responses from the private sector alone form the dense supportive web necessary to make a community whole in the face of disaster.

Corporations can be excellent partners for a recovery effort. However, there will always be a need for government, because there are simply too many interests that are not market-driven that are a focus of government, which correspond to the definition of public interest more than corporate or financial interest, and are deserving of protection for the body public. Corporations by definition serve the interests of shareholders, rather than the general public. The interests of the many in that instance may not correspond to the interests of a few shareholders. Some overlap may exist for companies that sell a broad array of products that are useful in disaster and stand to gain monetarily from full activation of a vast distribution and sales network (such as Walmart, Home Depot, or other entities with the resources, personnel, product, and distribution channels necessary to get product to a response/recovery operation and to those in need), but these are exceptions.

Government involvement does not equate with hostility to business interests. Businesses are made up of citizens; in addition to providing for employment, a business community might reasonably be reflective of shared beliefs. For example, companies should not endanger customers with shoddy or dangerous product manufacturing, or foul the air and soil in the interest of saving a few private dollars, when the result will invariably involve a call on government to clean up the mess with public resources. Business should not be unfairly targeted or limited by government. In practice governments and businesses have not had a strong track record in achieving this delicate balancing of public and private interests, and this has been to the detriment of citizens.

Businesses represent a power force in swaying the interests of government, while also employing most of the population. Because the relationship between the spheres of business and politics are close and dependent upon one another, it might be argued that the relationship they share leaves little room for the interests of the general public. Elhauge suggested, "government cannot be trusted to regulate in the public interest. Legislators are disproportionately influenced by organized interest groups and thus enact legislation enabling those groups to exact economic rents from others. Agencies tend to be captured by the firms they regulate and thus promulgate regulations to benefit those firms even though the regulations are inefficient and exploit consumers."[28] The ability of bureaucratic agencies to survive, politicians to be reelected, and businesses to have their interests heard might depend upon the stability each creates for the others through iron triangles.

Firms working together to engage in price fixing or bid rigging, to assure an outcome among themselves and therefore control of a public procurement process from outside the government, evidence an inability of firms to understand value in fairness and impartiality.[29] Contracts are for the taking in this view, and the idea of competition may be a convenient fiction that allows governments to engage in other than full-and-open competition for contracting without public outrage. Where this occurs, fairness and impartiality are words on a page and mere symbols, because they have no relevance to selection by merit and best value. This falls short of the ideal.

Public administrators can also act as interests within the system—this is another aspect of the system where neo-institutionalism has an explanatory, and perhaps even predictive, role to play. Purchasing agents may influence outcomes by what procurement process they select; the institution either checks that sort of behavior or makes it acceptable. Unclear language in a solicitation can lead to certain firms not being considered for a project, even though they may otherwise be qualified. An agent may choose to forego a new solicitation and extend a current agreement, which itself limits competition. "In environments full of restrictions, procurement of goods or services can necessitate securing separate approval from each bureaucratic layer, all of which function as monopolies concerned with their own rent-seeking and not at all with expediting consumers' passage through the maze."[30] This is another area where business and government do not see eye to eye.

Businesses have a role to play in the policy process, and are served by policy to varying degrees, particularly in the name of economic development. Businesses, like other organizations, are social entities and tend to communicate with and rely upon one another. In the context of disaster, this takes on a different hue—strong business networks might reduce vulnerability, which is a distinct positive.

On the other hand, negatives associated with a too-closely-knit business community are obvious. Business leaders and political leaders, and their relationships that tend to be less than aboveboard, are constant fodder for those that consider such instances as examples of the downfall of modern American public service. "Everybody agrees: bribery is illegal, unethical, and should be eliminated," but because politicians do not generally promise to do something "direct, immediate, and explicit"[31] in exchange for cash, these donations are not considered bribes. They suggest that we should call campaign donations "gifts" because they "create a feeling of obligation."[32] From that perspective, a general feeling of obligation may be even worse than a bribe, because there is no end to the quid pro quo. A politician may upon taking the gift become beholden to an interest group or a business, and expected to meet their ends lest there be significant repercussions come election time. In this view, business is something to be controlled and regulated because it cannot be counted on to look after the public interest.

There is question, though, as to whether close relationships and networks in the business community are truly negative. Businesses form networks, work with each other on projects, and buy products and services for one another. Employees live and work in a community. Taking an antagonistic view toward business does not solve the problems of government and it hurts individuals; there is a fine line between regulation of businesses to prevent abuse and paralysis of a business community through ineffective or unnecessarily complex policy schemes.

In disasters, politicians and government actors, the public, scientists representing various fields, and businesses may all have some role to play in managing, ameliorating, or contributing to the events surrounding the occurrence. Governments implement emergency management systems and provide for staff to address the needs of the public in the event of disaster, from first-responder concerns having to do with the protection of life and property, to long-term decisions that affect a community's ability to recover, such as rebuilding the local economy, dealing with housing displacement, covering expenses related to reconstruction, and other matters. The public wants to be protected, and when disaster strikes, it expects government to return the community to normalcy as soon as possible. Businesses are adversely impacted by disasters as much as the general population and government, and similarly want matters to return to normal so that they can conduct business as they had pre-disaster.[33]

Many communities plan for post-disaster redevelopment on a pre-event basis, so that the discussions between local government, community partners, and businesses that need to occur have already begun before a hazard event occurs. These discussions can assure a common vision for how a community can respond to a disaster event and, should such an event occur, what the process might look like for response/recovery and long-term planning to help attain a new normal close to the community's optimal resilience level.[34] Planning can also help protect a community's sense of place, by preventing gentrification and other community shifts that may lead to increased vulnerability. As Weick noted, planning is important but not primarily for the plans that may result—the interaction that induces conversations among stakeholders is most important. The organization benefits from planning because even if a plan is never used or never fully comes to fruition, the interaction still positively affects the organizations, its actors, and partnerships.[35] In disaster circumstances, planning for redevelopment will not only allow these conversations, but also attract other stakeholders to the table, while signaling to the community that those involved are serious about protecting the public.

## LOCAL IDENTITY: DISASTERS, GENTRIFICATION, AND PROTECTING A SENSE OF PLACE

Gentrification is the "restoration and upgrading of deteriorated urban property by middle-class or affluent people, often resulting in displacement of lower-income people"; this *American Heritage Dictionary* definition summarizes well the basic premise, including the displacement component. What it leaves out is the root of gentrification, which is something akin to globalization.[36] Along with the displacement of the people, indigenous culture and heritage are also potentially displaced or lost altogether in favor of the new conception of the place.[37]

The potential for gentrification frames the aftermath of disaster as an opportunity to insist on the uniqueness of a place, and protect citizens of all socioeconomic strata from the wealthy to the vulnerable, or to reimagine a community in a way that does not protect all equally. This would leave displacement of vulnerable populations, either purposely or as an unintended consequence, a very real possibility.

Gentrification is an issue of increasing importance given the tendency of many cities to support community redevelopment efforts that fundamentally change the culture of an area. The focus of such efforts is frequently downtown urban areas. Coupled with the nature of disaster and the rapid redevelopment that occurs during the recovery phase, the idea of a gentrified inner city may not receive adequate opportunity for public discussion prior to being implemented by local governments, through their direct action or their inaction in allowing others, such as federal agencies and others at that level, to make such decisions for them.

Both globalization and its local community cousin gentrification involve a "new colonialism" of sorts. There is a homogenization and universalization that is hostile to local culture and tradition that works in both. It would be impossible to ignore that the prevailing group in gentrification (as with globalization) is most often a "privileging whiteness," as Atkinson and Bridge put it: "In fact not only are the new middle-class gentrifiers predominately white but the aesthetic and cultural aspects of the process assert a white Anglo appropriation of urban space and urban history."[38]

"The gentrification process itself, which in different ways in different places deploys a wide array of artifacts and images to invoke deeply romanticized versions of history, within which issues of social identity and social status are often central. In so doing, gentrification transforms history into heritage, and untamed urban wilderness into domesticated urban landscapes."[39] The "new" of gentrification is not necessarily better, just more likely to appeal to the people with the means to live in the reimagined community. The new community is as real as people believe it is.[40] This approach to the concept is not unlike that offered by Benedict Anderson for defining the nation-state: a significant number of people believe that they have formed a nation, or a community, in this case, and therefore, imitating other places that have acted similarly, it is defined as they see it.[41] With gentrification, the old community, its culture and values, may be replaced completely by a new community, defined consciously or unconsciously by new residents.[42]

Whatever the new gentrification results in, it is very likely that it has lost the culture of the original place. The conception of gentrified place can be plain and undistinguished, even if it is safe and posh, in an effort to appeal to a broad base of potential customers without offending anyone's sensibilities. Urban transformation, such as gentrification or themed redevelopment, results in the dismantling of "older urban solidarities," which are then replaced with consumption spaces characterized by pastiche and appropria-tion.[43] In this view, the new cultural logic of global capitalism simply works to estrange "real culture" from its more authentic, more localized, origins. This now homeless vernacular is transformed into what Boyer referred to as "the spectacle of history made false."[44]

Gentrification typically finds an opportunity to assert such ends in an inner city under normal conditions, as a result of standard redevelopment efforts which have become a hallmark of cleaning up the urban core of not only the American metropolis, but cities around the world; unfortu-nately, means employed through redevelopment can destroy "communi-ties of color," while not providing the promised "social betterment."[45] Common means of downtown redevelopment have not had a great track record: "despite three decades of continuous redevelopment policies and projects, most American downtowns still have serious economic problems and are perceived, particularly by suburbanites, as inconvenient, obsolete, and even dangerous places. Hence some critics would declare that down-town redevelopment policies have failed."[46] A theme park–like, globalized downtown area in the typical redevelopment approach retains nothing of a historic downtown, and invariably the people who presently live in the area targeted for redevelopment are forced out, either overtly or because they cannot afford the new increased cost of living there. And yet, many local governments still believe that standard-issue, gentrification-ready approaches to redevelopment are the way to proceed; these methods might primarily serve to make developers wealthy at significant cost to the soul of a community.

Gentrification has become more cumbersome of late because the media has shown a light on it. Disaster scenarios can become an opportunity for policymakers to remake communities and cities in ways that fail to preserve culture, history, and civic identity, while displacing disproportionately vul-nerable residents. Local governments must guard against this tendency to be drawn into efforts to remake communities in ways that risk historical-cultural value while increasing vulnerability. The plight of racial groups in disaster situations has to be mentioned in the context of hurricanes and recovery, because the specter of Hurricane Katrina casts a long shadow on the idea of fairness and equity in recovery efforts. The New Orleans of today does not look like the New Orleans of even the early 2000s, pre-storm. The demo-graphics have changed. But our concerns are not limited to race, as the elderly and other vulnerable groups can also find themselves priced out of reimagined versions of their neighborhoods.

Gentrification itself can become a business, where community claims on belonging have little role to play in the final decision if left to the business world: "some business interests, intent on the [area's] 'revitalization,' have suggested that, given the value of inner-city land, policy would be better served if poor residents were simply relocated to peripheral areas, where land was cheap," even while "such claims ignore the collective constitution of the 'community,' and its moral right not only to continue as an entity, but to remain *in situ*."[47]

Gentrification also has an application to business, since business owners also are subject to the state of their surrounding environment. Business owners may live in a place and own a business in the same area. They may live in a residential area and own a business in an area zoned for a specific type of work. It is possible that business zones may become gentrified, in a manner similar to that seen by individuals or families in neighborhoods. Local businesses may be replaced by national companies, branches of conglomerates, or chain retailers, for example, in the wake of disaster. Some of this may depend directly on the action or inaction of local government, its stance on gentrification, and whether it will protect local small businesses.

## A CONCEPTUAL MODEL OF BUSINESS RESILIENCE IN DISASTER

Within the framework of community resilience, which is impacted by businesses, individuals, and the public sector, we explore business resilience more closely. In exploring variables that might impact business resilience, we are mindful of what the disaster and institutional literature suggests. To begin, disaster literature has taken up the idea that size of disaster may have an impact on business resilience. Size of disaster can mean aggregate impact of an event to a community, but it can also mean the impact of an event on an individual or a business. For example, a large event that has little impact on a small business will not be seen as a grave disaster event, but a small event with considerable impact will be seen as having severe impact by the business. For small businesses, personal finances can impact business recovery progress,[48] as can access to capital.[49] There is some evidence that smallness of a business can be a liability, making a business more vulnerable.[50] However, smallness may contribute to businesses opening more quickly, for family, pride, other reasons: In New Orleans after Katrina, for example, "sole proprietor businesses in less flooded areas [were] among the quickest to reopen, and . . . their reopening [did] significantly encourage neighbours to reopen."[51] Smallness could equal fleetness in response. There is a need to understand more about how small businesses react to hazard events.

Damage to supply chains and infrastructure affects product quality and standing of the firm in the marketplace, and can impact businesses' ability to rebound from disaster as a result.[52] Small or large, a firm may not stay in a community if infrastructure and supply chain have been severely disrupted; for large firms, options may exist to simply relocate resources to other locations, making a location within an affected community superfluous. From the literature on vulnerability, we might expect that firms owned by minority groups, or women-owned firms, may have greater vulnerability.[53] These firms having strong financial resources alone may have little to do with outcomes and whether they are truly less vulnerable than other similarly situated firms, all other variables being approximate, as disparity studies commissioned by local governments frequently suggest.

Pre- and post-disaster purchasing and economic development practices can be examined through theoretical lenses to gain additional understanding of both underlying motivations on the part of local government leadership and the role these leaders and their governments play, or fail to play, in the recovery process. The measures of months to return to normal and percentage of business capacity after disaster are indicators of recovery, which is tied to resilience. While a number of other factors might influence these indicators on an individual business level, they nevertheless serve as initial points of departure into a much broader subject—local government impacts on business resilience. Focusing on how government operations can influence business recovery and the ability to recover leads in part to community resilience. Concepts from vulnerability literature can play a role in understanding how vulnerability can affect resilience in the business community after disasters. Business resilience is impacted by local government policy, specifically in procurement and economic development, through direct institutional action and indirectly through the culture of institutions in the neo-institutional mindset. We make use of the following conceptual model for business resilience, as indicated in Figure 3.1. This model is revisited later in chapter 7 and expanded upon, in exploring how business resilience is interrelated with that of individuals and public institutions, leading to cumulative community resilience.

Given this initial premise, that institutions matter and have an impact on business resilience, and in turn community resilience, we turn our attention to the underlying activities for institutions that we will examine—economic development and procurement institutions in local governments. Both have theoretical underpinning that leads to their application and

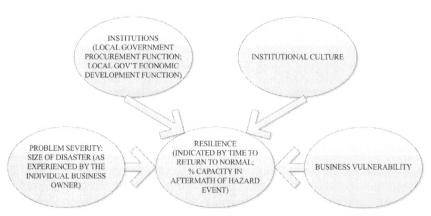

*Figure 3.1*   Conceptual model of business resilience.

insistence on implementation of a system of regulations as the rules with proper enforcement. Economic development is reflective of local culture, institutional values, and political tendencies, even when a public-private partnership is utilized to provide the service, or a private sector economic development organization performs the function. Purchasing is an institutional proxy on a number of levels—from fairness and approachability of a given unit of government to how the governmental entity interacts with businesses of all sizes. Like economic development functions, procurement operations say much about an institution's values.

# 4 Public Procurement and Economic Development at the Local Level

An examination of public procurement and economic development is valuable for our study; these functions are more unfamiliar than some areas of government, but nevertheless have influences on government and community. Both areas also have a basis in the rational province of local government to ensure the public good. While these two specialties are frequently ill-understood by the general population and prone to oversimplification or minimization, professionalization has contributed to legitimacy that is helpful in normal operations as well as in times of crisis. We examine each specialization individually in context of local government, and more specifically, how each functions in a hazard event response or disaster condition.

## PUBLIC PROCUREMENT

In providing information about how a public institution operates, the procurement function can be especially telling. While the idea of public procurement is seemingly uncomplicated, the expectations of the public sector for transparency and accountability heap complexity upon processes. Coupled with the nature of people within institutions, processes that are the same in two places may yield very different results, due to the intersection of people, procedures, and the institutions themselves. When a disaster strikes, unless institutions are primed to proceed steadily and with assurance, intricate processes under duress can lead to unintended or damaging outcomes.

Public procurement, as I use the term here, is government contracting at the federal, state, or local level, where government contracts directly with private suppliers of goods and services. Ideals of public procurement are thought to include open competition, impartiality, equity, and providing a product or service at the least possible cost to the customer, ultimately meaning the public.[1] Others propose a transition to sustainable procurement which considers "post-materialist concerns and social consequences such as quality of life issues that include equal opportunities and community impacts . . . This also means that for a growing number of vendors and suppliers, competitive advantage will become rooted in such new capabilities as pollution mitigation and prevention, eco-friendliness, social development, and stakeholder dialogue."[2]

"Public procurement has been a neglected area of academic education and . . . research interest by academicians."[3] While professional organizations have "worked to make the public procurement workforce more and more professional,"[4] the field has perhaps been less research-oriented. Resultant policies, even where they are standards-based, are often particular in practice;[5] this owes in part to the distinctive considerations, political and otherwise, of locales. Because public procurement leaves open the possibility of serving interests larger than simply the best price for a product or service, all the difficulties that plague public policy in implementation may show up in some form in public procurement, at the point where intent meets practical circumstances. Seeking to serve greater public interests through procurement activities is so common now that purchasing agents have begun to see that there is a role to play in purchasing beyond the best price for a given specification.[6]

Fairness and accountability to the public are central to an ideal vision of public procurement. For example, the Federal Acquisition Regulation of the United States Government has as its goal "to deliver on a timely basis the best-value product or service to the customer while maintaining the public's trust and fulfilling public policy objectives."[7] The system that surrounds the federal regulations governing procurement has a similar mindset: "All participants in the system are responsible for making acquisition decisions that deliver the best value product or service to the customer," but tends toward a more vague definition of best value, noting that "All participants in the System are responsible for making acquisition decisions that deliver the best value product or service to the customer. *Best value must be viewed from a broad perspective and is achieved by balancing the many competing interests in the System.* The result is a system which works better and costs less"[8] (emphasis mine). It has become increasingly clear that public procurement can have a role, positive or negative, in responding to the call of various public policy obligations by wielding the power of the government's checkbook.

However, government itself increases the valuation of certain interest groups' wants at the expense of others and arguably the public at large, through policy. As Wilson writes, "just as important and more pervasive in their effects are the legal constraints placed on the procurement process to insure that contracts are awarded "fairly"—that is, in ways that allow equal access to the bidding process by all firms and special access by politically significant ones. [The Federal Acquisition Regulations contain] dozens of provisions governing the need to give special attention to suppliers that are small businesses (especially a "small disadvantaged business"), women-owned small businesses, handicapped workers, or disabled and Vietnam-era veterans, or are located in areas with a labor surplus."[9] Fairness is defined, at least in public procurement, by interest groups. Some groups are given special consideration, while other groups are not, due primarily to political agenda setting.

We know more about public procurement through recent academic work, and there has been much standardization of practice among government agencies, but as a field of practice, procurement is a founding governmental role. American experience with government contracting extends to George Washington having difficulties with suppliers, who were trying to profit from the Revolutionary War,[10] and the first purchasing by federal government in the guise of the Continental Congress in 1778, an "arrangement [that] led to excessive costs and possibilities of fraud."[11]

Contracting for products and services is standard practice at all levels of government, but such reliance is particularly noteworthy in state and local governments,[12] where private entities might provide everything from road construction to staffing of public libraries. Contracting might lead to reduced cost for the same products and services, the use of experts who can develop better solutions for public problems than can regular government staff, avoidance of red tape, and flexibility.[13] It can also lead to capture of government through limited choice in contractors, a tendency toward over-specification which raises costs in describing the needs of government when soliciting for contract, underperformance, overregulation, the absence of incentive to remove poorly performing contractors, given the difficulties of working within a bureaucratic contracting system, and corruption.[14] It can be said that "corruption notwithstanding, the more contracting out, the more resources necessarily will be devoted to contract writing, specification of standards, performance monitoring, and auditing . . . In addition to providing a new outlet for red tape, another layer of accountability and oversight bureaucracy will be necessitated."[15]

Processes in public procurement have been described as multi-objective, meaning that besides price, ranking of responding vendors can also consider "quality, quantity, delivery, performance, capacity, communication, service, geographical location," among other factors.[16] Procedures and transparency are important, but fuzziness in ranking can result from the sometimes conflicting criteria.[17] The prognosis can become so complex that local governments using existing contracts from other locales or the state can be considered a best practice, because such *piggybacking* reduces time from identification of the need to receiving the product or service. It also potentially prevents firms that might be interested in a government contracting from ever obtaining one, in that competition is reduced through fewer opportunities to bid.

It is sometimes not enough for firms to compete for projects, even when they might win projects on the basis of merit given their experience and expertise. Attempts to influence public officials may result in conflict of interest concerns, where "government employees steer business to firms that have been in the past—or may be in the future—their business connections."[18] As governments contract more projects out with the intent to save money over services provided by the government itself, the prospect of corruption increases, leading some to conclude that one "make[s] more money rigging bids than robbing banks."[19]

This brings us back closely to the vague nuancing that menaced the goals of public procurement. Even with strictly worded statutes meant to cause government business to be conducted "above reproach and with impartiality,"[20] and directives against receiving compensation less a project be canceled[21] or worse, we nevertheless see that seeming violations of the letter and spirit of such prohibitions are not uncommon in government procurement. The multi-objective nature of the public purchasing leads to practical problems.

Government contracting, to meet the standards of fairness and high quality public service, requires "a sophisticated collection of different administrative tools . . . tailored to the special implementation problems of contracting."[22] Essentially, complex regulations and processes need to be drafted and enforced, assuming that contractors and government officials will violate them for self-interested reasons. Government contracting "also requires incorruptible and highly competent government officials to manage the contracts."[23] This is an elevated standard even for well-intentioned public servants. In an institution with a culture that discourages proactive responsibility and accountability, it may not be possible to attain such a standard. Procurement is an intricate assignment with extensive requirements, necessitating knowledge and skill among its specialists; professionalization is less an option than an expectation.

Procurement in the public sector, Kovács has suggested, should be fair and impartial, consistent, transparent, efficient, economically-minded, well-documented, and accountable.[24] Buying goods and services in government, using public money, is more complicated than private purchasing, and this is reflected in the list of expectations. The cynical observer might have trouble believing that government procurement is ever fair, with the presence and constant influence of interest groups, each with its own agenda, working to force the outcome of a public process to something other than a fair and impartial end. For instance, it is possible to make a connection between political donations and contract awards, which raises significant questions not only for fairness of purchasing processes, but for fundraising in election cycles and use of power.[25] Agreement on founding principles of accountability and fairness aside, accomplishing such a standard is a lofty goal.

Consistent with ideas of supporting the general welfare of citizens, "Societies are just . . . to the extent that their major institutions conform to principles of need, desert, and equality."[26] For procurement, those approaching government with products and services to sell ought to have an equal chance of selling their products and services. They should, in other words, have an equal right to the pursuit of their happiness, through the pursuit of their business endeavors. Giving equal, merit-based chances to gain contracts from government allows the governmental entity to make available to the public those resources gathered from taxes and other sources of state wealth, in the cause of building a stronger society. It is just and morally right to do so. The decision to award projects to certain vendors should not rest with who those vendors know, or their connections to the politically powerful, when the merit of a solicitation response would indicate a different award decision in procurement.

This is one point where social justice proponents and pure-market critics of governmental involvement in leveling playing fields should agree—that outside influence should be limited. The intervention on grounds of social justice by government is a Rawlsian concept—where the normal system does not seek to improve the position of those worse off, it is right for government to become involved through some assertive action. The needs of those who do not have should not necessarily be secondary to the wants of others, just because the system has previously allowed that result.[27]

In disaster situations, competition and fairness in public procurement are perhaps even weightier considerations given the possible impact to a local business community. When reviewing procurement codes in practice, it is interesting to note how they differ from this central tenet. For example, the Federal Acquisition Regulations clearly state what is meant by fairness in procurement operations and list requirements for full and open competition, but the additional detail provides exceptions for "unusual and compelling urgency" and "public interest."[28] To some extent, this involves a value judgment that can unacceptably abandon local interests.

Procurement systems operate in political and social environments and are influenced by them.[29] The using agency or department—the agency that is tasked with the project work in a given instance—identifies a need for the project, or is given an assignment that falls under their typical scope of service as an agency. A project manager is identified, who must then determine, with the assistance of the procurement department, the best way of approaching the solicitation for the project, given the total value, the type of work, service, or commodity, and the scope of service—specifically whether or not the solution to the problem being posed by the solicitation is well-defined. Specifications are drafted meeting general requirements, any specific obligations are noted as far as local requirements for the solicitation, and the project is bid and awarded.

This type of project solicitation and award process is useful for explaining generally how government procurement works in broad terms, but in reality, government procurement and the procurement codes that are enforce constitute a dense thicket of regulations, procedures, deadlines, and even bureaucratic stasis. The often paper-based processes to buy even simple products can be inordinately slow relative to the complexity of the product. E-procurement approaches, which can speed up processes and make purchasing operations more transparent, are not universal.[30] Where the competitiveness that is the goal of public procurement takes a secondary role compared with political influences, bureaucratic responses and the inability to assure fair outcomes can lead to public outrage. Instances of public corruption are a point of concern for many governments, and procurement operations may be a flashpoint for such concerns.

Competing interests exist within the procurement process as well. The project manager is perhaps the closest staff member available to a "subject matter expert" outside the staff of the firms that will be responding to the solicitation in an attempt to win the award. One relies on this person's knowledge as a municipal representative to help insure that the public receives a good deal for its money. Evidencing administration's multifaceted calling, this person's role is to act as a sort of impartial project-level judge, knowing the facts and helping the county navigate the selection. Yet, the person is not elected and is only accountable to the public through eventual accountability, through perhaps several supervisory layers to the county administrator or city manager.

A project manager may want to have as many eligible prime contractors or prime consultants respond to the solicitation for his or her project as possible, because having more options would, at least in theory, provide for greater competition and a better outcome for the county or municipality in terms of price and quality. More options might lead to a better final choice. Another option is that the project manager will over-specify the project to limit the pool to just a few preferred respondents. In some cases, project managers may influence choices that are made with regard to the solicitation—including bundling together smaller projects into a larger project, that is effectively out of reach of smaller firms.[31]

The purchasing official is in a powerful position to influence the outcome of a project. A purchasing agent, properly motivated, can shepherd a project through even a cumbersome process in record time, turning the awarded contract back over to the project manager for contract administration, and do so fairly. A purchasing agent can just as easily recommend an approach to solicitation that is more difficult than needed, not make the appropriate specifications, send out a solicitation document that requires amendment, and therefore extension, and make similar errors that can lengthen a procurement process by months and even years. Purchasing agent actions can amount to anticompetitive behavior, whether they involve unreasonable requirements, insufficient time to respond, or the provision of insider information to preferred firms.[32]

As an administrative function, public procurement includes judicial-like proceedings, through the use of protest hearings; legislative-like processes, through the drafting of procurement code in initial form, which has the function of law in practice; and executive-like activity, through the award authority given to purchasers to encumber governments up to a certain amount with the general grant of an elected body. Like project managers, government purchasing agents are not elected and yet they are popularly required to be as accountable, if not more accountable through their often at-will employment status, than elected officials. The learning environment in public purchasing is tough and unforgiving. Expertise and a broad knowledge base, informed by experiences in a variety of solicitation contexts and for many different products and services, are the only sure way to achieve success professionally. The political and public expectations weigh against sound practices and make the tasks at hand more arduous.

Public procurement has recently seen a marked movement toward the use of technology—from vendor databases to fully electronic, cloud-based procurement systems facilitated by third parties. This tendency can be seen as the most recent extension of the field's inherent technical-instrumental rationality. Technical rationality refers to "a way of thinking and living that emphasizes the scientific-analytic mind-set and the belief in technological progress."[33] Technical rationality grew out of a general desire for efficiency and productivity at the height of the industrial revolution. The concept suggests that as technology moves forward, people can expect improved standards of living and greater precision in the activities of government, all for a lower price. Further, the problems and complexity of government can be overcome through an approach that stresses rationality and analysis. This has fostered the belief that systems are needed in the public sector which will allow the public sphere to function effectively; these systems will allow government to succeed in its aims even with financial and staffing limitations.

Relatively more than other agencies that serve as the engine room of modern government, it could be said that public purchasing has acquired a veneer of instrumental and/or technical rationality out of need. Modernity has deified instrumental rationality as an enabling logic. The hierarchy authorizes the actions; the professionals insure that technically rational knowledge is appropriately applied; and the technology facilitates the efficient and effective disposition of one's duties. Procurement has a need to be seen as a solver of problems; as the acquirer of products and services in an organization, the agency is a natural power center, but often hides under the pretense that it is merely a means to an end, or a value-neutral proposition.

There may be an overreliance on the role of technology in assuring quality in public service. A system that takes more information into consideration is not a replacement for responsiveness in government or how approachable government is to citizens. Replacing people with systems may lead government to a false sense of security, in that there would be an increasing and misplaced sense of comfort in knowledge and ability to respond quickly to events, and questionable representativeness in the response.

As Kenneth Hewitt suggested, "uncertainty, at once defines an empirical and practical dilemma, excuses the failure to control, and justifies expert systems as alone able to probe and domesticate these further reaches of environmental and social wildness."[34] The belief that the imposition of systems reduces uncertainty, and that the reduction of uncertainty makes society safer, is a prevailing, though most probably mistaken, assumption. Approaches that stress the role of technology in a disaster scenario over human roles can provide a glossy veneer over an otherwise inferior construct of relationships and functions that will nonetheless fail when strained.

As public procurement continues to grow as a field, it must come to terms with the concept that it is not only a means for government. Instrumental rationality has the capability to be constructive or destructive, but is frequently destructive because of the limited nature of rationality and the self-serving approach of individuals. Tied up in so many systems and processes, and with rules of all kinds to provide backing for any possible outcome, public purchasing may overwhelm the very public it is supposed to serve and use process as an excuse. In some public purchasing contexts, only a few firms are ever truly successful in winning many contract awards of considerable value; they have often achieved this success by engaging the services of lobbyists and attorneys, undermining competitors through political avenues, and playing to their strengths while taking advantage of the weaknesses in procurement processes and solicitation documents. Firms may underbid a project when they plan to make up the supposed loss with change orders; other firms take on small businesses as subcontractors without any real intent to use them on the project. Such gamesmanship speaks of a disconnect between policy intent and practice. In erecting ziggurats of procedure, purchasing agents are sometimes as much victims of process as implementers.

Public procurement, like many government programs, may exist as a code or set of regulations that provide at least the façade of rationality and lack of discretion in contract awards. This we might equate to fairness through objectivity, but we should look more closely, as many considerations—such as how to procure a product—are subjective. Procurement systems may assure accountability to the public at large, and they may deflect criticism when difficult decisions need to be made and a governmental entity needs to be defended in its decisions. How those decisions are made is part of the culture of an organization, and official actors, in their interaction with the system, are participants in the creation and reification of a culture. The responsiveness of this system to the needs of the public is critical to understanding resilience because it speaks to the values of the wider institution as well as those of the community.

For example, electronic procurement systems have received a great deal of attention in providing for faster, more transparent, and generally improved responses to the need for government purchasing. However, there can be unintended consequences to using such systems. First, using an e-procurement system requires a skill set that not all business owners have in equal measure, even though they may be competent professionals in other respects. Placing a solicitation on an Internet website does not necessarily mean that more firms will respond to it; quite the contrary, where traditional approaches might require that an agent find no fewer than three responsive and responsible bids, website responses may stimulate extremely limited response from only a few bidders of varying quality. An over-specified solicitation or one where would-be bidders have questions may draw no interest at all. In other words, technology cannot remedy poor procurement practice. Second, e-procurement expands the reach of a local government purchasing operation worldwide. Local businesses that previously may have had much success with city or county contracting opportunities now must compete against businesses from other states or countries. While this may result in lower prices, it may also result in a crowding out of local businesses, which has detrimental impacts to local economies in quantifiable ways. Third, systems are provisioned by third parties which may not be as accountable as the governments themselves. Firms may have to pay some kind of fee to even become registered as vendors—and this is before they even respond to a solicitation or win a contract.

Procurement systems rely on solicitation notifications that are sent to registered vendors. This can create an insulated pool of vendors and decrease likelihood that new vendors will have a chance to become involved in public procurement. Many potential vendors are not aware of contract opportunities in a way that makes them equally able to compete and win projects, since the system is the primary mode of contact. Even if information is available through a website, the website may be difficult to find or navigate; information available through traditional print sources may be overlooked. Direct contact on a timely basis to potentially interested firms works best, but only the firms that are registered receive this benefit.

The technical capacity divide still exists and is notable in small businesses, which do not have in-house information technology departments, groups to prepare solicitation responses, lawyers, or financial resources to hire lobbyists. Small businesses may have a learning curve in approaching public-sector procurement which could result in them being unsuccessful in responding to bids for months or even years, given the infrequency of solicitations for some commodities or trades. There will always be a finite number of project opportunities available, and size of business carries distinct advantages in public procurement. The system's reliance on instrumental rationality provides distraction from the fact that it is difficult for some businesses to maneuver, and that it leads to different outcomes depending on the capacity of businesses to approach and make sense of it.

During economic downturns or crises, public procurement could be even more of an indicator about how an institution processes policy—how it constrains actors and plays by sets of rules. A rule of thumb in emergency management is that practice is essential—what institutions practice becomes engrained and second-nature. The time to practice successful behaviors is not when a disaster is occurring, and the process-driven approach of public procurement is no different. The tendencies in place before the hazard event tend to be amplified by the event; if the institution and its staff were mostly accountable, efficient, and process-oriented, this may continue and serve the community positively in response and recovery. Consistency in process and documentation of purchases, for example, can be helpful in obtaining reimbursements from other levels of government. Even in a disaster situation, there is still a need to try to get the best price possible. On the other hand, a community that had a procurement process that did not have adequate controls in place, or where documentation was limited, may see these flaws turn into failures on a grand scale.[35]

Local governments benefit from opening up to the local business community well in advance of any hazard event, and learning more about what the business community can provide. If electronic procurement systems are used, a lack of businesses in needed specialties as registered vendors may create difficulties in responding and slowed response and recovery. Local businesses are close in proximity and potentially able to deliver products or services quickly, and there is the added benefit of an economic multiplier, where spending money in one's own community continues to benefit the community. It might go without saying that it is smart to shop in one's own town, but in times of disaster this is a crucial point; the local firms may need the city or county government's business critically in order to fully return to operation and attain a sense of their "new normal." When supply chains and customer bases have been disrupted, local government spending in the local market, by a mindful purchasing group, can make all the difference. Technical rationality aside, some entities have found that soliciting through the comprehensive technology solution pales next to opening the community phone directory during a crisis.

## ECONOMIC DEVELOPMENT AS A ROLE AND EXPECTATION

There have been attempts to define economic development over the last century as the field has grown into a profession, but a consistent definition of the field between contexts and applications has proved elusive. In the early 1970s, a definition of development offered by Oberle, Stowers, and Darby links ethics, sociology, and economics to increasing "life chances"; this connects behavior to collective utility.[36] What this definition lacks in recency, it perhaps makes up in prescience; economic development continues to raise these same issues in its present definitional discussions, as noted in a review of definitions offered by Blanco.[37] Economic development includes policies to encourage "price stability, high employment, and sustainable growth," as well as programs to provide for infrastructure, which has broad importance to economic development, and to improve the "business climate, through . . . business finance, marketing, neighborhood development, small business development, business retention and expansion," and other activities.[38] Economic development activities can be carried out by public organizations, private organizations, and public-private partnerships. It may ultimately matter little what kind of organization carries out such activities given the close relationships between local government and business at the local level; whether entities are public, private, or a combination, activities in this area are highly reflective of institutional and community values.

Partridge and Rickman argue that "economic development in a region should be equated with utility of its current residents."[39] Job creation and retention, enhancement of tax rolls, quality of life, and a host of other reasons all encourage governments to engage in economic development. How one entity decides which economic development measures to employ is dependent on the region's individual characteristics, as not all measures work equally well in all places. There is variability in the package of which economic development initiatives are recommended given a set of circumstances.[40] The field in practice involves both application of theoretical principles and, in the hands of especially capable practitioners, a fair amount of art.

Institutional and community context influences the choice of local governments to engage in economic development pursuits. Economic development programs promise great benefits to governments that consider enacting them, but not every city or county has such programs. Changes over time in how government does business led in some cases to relatively less commitment from public officials and bureaucrats to these programs, and as a result less reason to oblige the desires of constituents concerned with development. Where a development mind-set existed absent public-sector involvement, efforts were led by community groups and the businesses themselves. In such an environment, unpopular policy decisions would perhaps not have the same repercussions as they would in a place where much more counted on the policy choices made by government, even if government partially funded the private-sector initiatives. In more public-centered locales, by comparison, government officials would be relatively more approachable, and perhaps even more held to account, by the public, and would have more of a reason to discern public desires for development and be responsive to them.[41]

Where economic development policies exist, there is a need for focus internally and externally—internally for enforcement of the policies and procedures enacted to support such programs, and externally to attract, retain, and expand businesses. "Economic development occurs when local leaders choose to identify, invest in, and develop their own set of comparative advantages to enable workers, firms, farms, and industries to better compete in regional, national, and international markets."[42] Economic development is both a choice and an investment, and requires consistency in vision and application to see results. A pro-business orientation, in order to be effective, must have behind it significant political will at a high organization level, perhaps from a senior official acting as a "political entrepreneur . . . [gathering and risking] political capital in support of development."[43] There must be recognition among political leadership that the cycles and interactions at work in economic development theory actually play a role in the future of communities. A good example of a city pursuing an entrepreneurial political economy would be Dallas, Texas, from the 1930s to the 1970s.[44] In other locations, a negative cycle must first be broken. Government efforts must be focused to break a vicious cycle of poverty, for example, which can lead to more poverty on supply and demand sides.[45]

Economic development efforts tend to thrive in places that are stable, and where the belief exists that economic development policies contribute to this stability. This would include places with strong local elected officials, with a vision and considerable authority for motivating the institution to engage in such development efforts; low turnover and a tendency to support government processes, rather than pursue privatization as an ideological goal, are also frequently evident. A strong mayor who wields considerable political power may be able to support economic development policy generally and to be more forcefully involved in project-specific endeavors.[46] A government that has long-term employees and plans for succession, and does not experience high turnover, will be less likely to have lapses of institutional memory,[47] which could, for instance, adversely impact the durability of an economic development initiative.

Where an economic need exists, such as unemployment, there is a necessity to actively address, or at least seem to address, that need, for economic as well as political reasons. Areas that are undergoing economic stress may want to try some kind of economic development program.[48] Not everyone agrees—a study by Martin Saiz, for example, suggests instead that governments tend to expand economic development efforts as unemployment lessens, which is somewhat counterintuitive; further, recruitment activities of competing areas may actually be driving much economic development activity.[49]

Program selection has to do with the institutional reality of the policy setting. Where local executive power is considerable, there may be more opportunity to institute negotiation-based programs like tax abatement structures to attract businesses. In places where this is not characteristic of the institution, one may see business assistance tools used, where management skill alone may make the program successful. Programs, depending on the strength of the institutions that guide them, may have more or less necessity and support, which leads not only to the choice of policy instruments, but also to their efficacy once instituted. Even the selection of zoning laws, and the motivation behind such selections, says something about the institution and what a governmental entity is willing to do to spur or constrain the development to bring about the community it visions; land use planning affects physical space and creates "a metaphysical place [through property rights] that further constrains future choices."[50] Institutions are also externally constrained, by federal or other law, which can in turn affect potential initiatives at the local level.[51]

There is disagreement as to what types of economic development approaches work best; just as one example, growth of governmental responses to economic development and intervention in markets may have a detrimental effect by discouraging private-sector growth.[52] Where a certainty exists about "the right thing to do," then perhaps that certainty is misplaced, even when it is acted upon and made law. This undermines the idea that rational thought carries the day on a decision more than institutional norms of appropriateness and the constraint that extends from past experience and institutional knowledge.

The decisions being made in development institutions are themselves constrained by previous decisions—such program choices are "path dependent," and undoing the past can be exceedingly difficult institutionally. Path dependence is limiting for both individual and organizational decision making.[53] If the decision is made to change the direction of the institution or try something else, the costs of doing so can be extensive. Changing the direction of the institution from a policy perspective can be a trying task.

Institutions intervene in the community's affairs as they articulate group preferences, as they do in enacting economic development policy. The choice to support economic development initiatives, decided upon by elected officials and carried out by government, is an articulation of that desired outcome on the part of the community.

Economic development as a practitioner field favors corporate visitations and efforts to keep businesses in a community—what are collectively referred to as business retention efforts. These can occur through either public or private auspices, although there is something to be said for the peer pressure of other business leaders for a business owner considering relocation, stressing a business's individual importance and contribution to the collective strength of a business community. There is some agreement that the public-private partnership approach can work in local government and be successful in achieving growth for the local business community, because the public and private partners have different means available to them. Government has access to elected leaders and other agencies, such as permitting and licensing agencies. Private-sector groups perhaps have more immediate access to financing entities and are more capable of creating networks among the businesses themselves.

As the professionalization of economic development has increased, it has become more expected that those engaged in this work will play a part in both response and recovery when disaster threatens. Economic developers play a variety of roles in disaster situations, including: determining how hazards pose a threat to specific industries in a community, conducting a benefit-cost analysis on potential recovery programs, working with small businesses to discuss potential vulnerabilities and risks, communicating information to businesses, politicians, community partners, and others about how to reduce risk and, after the event, recover more quickly, applying for recovery grants, coordinating a network of business recovery centers (BRCs) with public and private partners, and many other tasks.[54]

Economic developers can lead business recovery emergency support functions, including the aforementioned BRC locations. These locations can be staffed by partners from local, state, and federal levels, and from the private sector (especially financial institutions) to assist businesses with completing assistance paperwork quickly, addressing individual issues, and getting businesses back up and running. BRCs can be a critical point of contact with the business community, linking economic development staff with a business recovery liaison at a local emergency operations center, who can in turn resolve business-related issues directly with other emergency support function groups. They might be located at community centers or libraries, or under unfavorable circumstances, wherever it is possible to locate and serve business owners. Resourcefulness, clarity in services provided, and staffing support are keys to BRC success. It might be possible to create a virtual business emergency operations center, using websites, text messaging, and other communication channels, to support notifications and knowledge transfer prior to, during, and after any hazard events. Economic developers' role in pre- and post-disaster periods is to form the core of a business response network, to allow recovery to begin.

Reaching out to small businesses in particular is imperative—to not only raise awareness of threats, but to plan for how small businesses will respond to them. Speaking outside the perspective of disasters, small businesses are a major focus for economic development. Entrepreneurship and working to foster growth in small businesses, and retain businesses in a community, are core economic development functions.

## SMALL BUSINESS PROGRAMS

Governments have a variety of reasons for intervening in the market to assist small businesses (small and medium-sized enterprises, or SMEs)—from economic development to sustainability, to "fighting social exclusion through a contribution to regenerating urban areas."[55] The social significance of the success of a small firm—popularly portrayed as a sole proprietor supporting a family—is an expression of the common dream of being one's own boss. The subject is not terribly technical for policy purposes, and is unique and interesting to broad segments of the population because it cuts across firms of all specialties and skills.

There are a number of reasons local governments engage in efforts to support small businesses, among them job creation and innovation. However, new businesses are also a source of volatility, and ways of counting new jobs, such as net versus gross new jobs, confound our ability to calculate the real effect of small businesses relative to large businesses.[56] Some have suggested that policy might focus on entrepreneurship and give due regard to the age of firms in addition to size.[57] Others suggest that small business policy can be an important part of job creation, but efforts should be targeted toward those firms most likely to succeed and grow.[58] In small business development programs, we may see options for government intervention ranging from attempts to insure equality of opportunity (leveling the playing field) to forms of distributive justice that attempt to rectify past discrimination or simply serve to spread around the wealth.

Local governments find reason, beyond providing equality of opportunity for contracts, to support small business initiatives. Of primary importance is the role of such programs as economic development instruments.[59] The small business program is a tool, beyond opening up contract opportunities to small businesses that would otherwise be edged out by large businesses, which serves economic development ends. Cities and counties that actively do business with small businesses keep public money in the local economy,[60] where it is spent by locally employed people; this helps to keep businesses operating so that they can maintain and add to the employment base, while encouraging growth and innovation in the local economy. Such a program may accomplish this through direct contracting or means of subcontracting—for example, establishing a percentage goal on a project representing work identified for targeted small businesses, to be addressed by a prime contractor in its response to a solicitation.

As emphasis shifts away from minority- and women-owned business programs as treatments of institutional disparity in contract practices, toward race- and gender-neutral alternatives, the programs become more like economic development instruments and less like means of addressing past institutional discrimination. Local small business programs typically require that certain contracts be limited for competition to small businesses deemed eligible by the local government, for economic, size, and/or geographic reasons; they are either solicitation requirements or items to address for points in evaluation of responses.

One question in analyzing the business development program model is to ask whether politicians have any genuine desire to see these programs work. If they act as symbols only, the regulations might be on the books but the formal or informal culture could deem implementing the regulations as inappropriate and resist their enforcement. Those that would seek to implement them might find that the institution would attempt to constrain such efforts through bureaucratic inertia, if not outright hostility.

Further, it is possible that these programs may have undesirable, though unintended, consequences. Noon found that race-based programs are frequently criticized for "corruption, lack of government monitoring, confusion over implementation of the program, and fronting."[61] However, referring to an example program, "the most frequent criticism is that the BEE [Black Economic Empowerment] program's reforms have benefited the most politically connected and wealthy blacks at the expense of small- and medium-sized black businesses. Procurement contracts have largely gone to large black-owned businesses, despite a primary goal of the racial preference program being to end the domination of large corporations in government contracts."[62] Such criticism suggests substitution of one form of discrimination (race) for another (wealth), and fails to solve the problems originally targeted.

Small business programs and initiative to benefit small businesses undoubtedly have a strong symbolic component. Small businesses form the bulk of all businesses, and in terms of employment, employ a large percentage of all employees as a group. The number of firms that are small would make any small business program extremely competitive; the symbolic effect of the program is that government works to include small businesses, even though the sheer number of small businesses would tend to limit the availability of an opportunity to any one business. Coupled with political considerations and the difficulty of approaching public procurement opportunities, small businesses eligible for such programs may have little opportunity to become any more competitive than they would without a program. Depending on one's perspective, the purpose of the program may have been served, if the effect is to show government's support of small business to the public. There is therefore a need to evaluate such programs more closely in terms of performance in real terms beyond the obvious symbolic or political benefits. Small business programs may serve a positive societal interest and benefit businesses, but only if they are run fairly. Otherwise the policy may simply provide a pretense for outcomes that would have occurred regardless.

Professionalism is needed among administrators in this area just as it is needed in purchasing, because programs that are not operated with a vigilance to protect fairness may yield outcomes that are easily manipulated. These programs are also in need of close examination of their benefit for cost, beyond the symbolic nature of providing opportunity to small businesses.[63]

## PROFESSIONALIZATION AND LEGITIMACY IN PUBLIC PROCUREMENT AND ECONOMIC DEVELOPMENT

Louis Brandeis wrote the following of professions:

> *A profession is an occupation for which the necessary preliminary training is intellectual in character, involving knowledge and to some extent learning, as distinguished from mere skill; second, it is an occupation which is pursued largely for others and not merely for one's self; third, it is an occupation in which the amount of financial returns is not the accepted measure of success.*[64]

Professionalism is an important question for government, particularly in an age that is increasingly distrustful of government and suspicious of its interventions in individual lives. The public frequently gives government agencies no pass—with government officials portrayed as lazy, incompetent, or corrupt. The portrayal is, for the majority of public-sector employees, exceedingly demoralizing and unfair.

Public administrators are called to do more than simply implement policy enacted by elected lawmakers. Though unelected, they engage in rulemaking and creation of administrative procedures to carry out policies and programs, under the authority of enacted law, and are held to account through public participation in such activities and oversight by all three branches of government. Public administration has long sought to frame the role of bureaucracy as bringing competence and expertise to questions of grave importance. Professionalization is "the differentiation of a practitioner from her social environment."[65] Professionalization contributes to legitimacy in the public sphere.

Advocates of public procurement and economic development have sought to professionalize the practice of both fields. Technical rationality gave rise to the idea of expertise residing in public administration, for the proper conduct of governance, and it follows that public agencies might, after some period of growth and development, arrive at the development of a professional consciousness. But what of government purchasing, which might seem on its surface as merely a means to some other end, rather than a valuable role in itself? Or the work of economic development, which aims to increase employment and investment in communities—something that has considerable private-sector benefits and a less-clear rationale for public return on investment, particularly in incentives programs?[66]

The concept of professionalization in public service can be applied to both public procurement and economic development, at all levels of government. Neither public procurement nor economic development has direct correspondence to a specific academic field. Practitioners are likely to be generalists representing a broad range of both education and expertise. Both specializations focus on work that may be thought to be fairly straightforward, but that is upon closer examination complex and potentially subject to both error and corruption. There are legal issues that must be addressed and understood, and often complicated procedures to be followed. While being a generalist with broad knowledge is helpful in approaching these fields, only expert knowledge of a detailed and applied sort will allow success for one seeking growth in either purchasing or development.

Training is therefore a necessity for staffers entering these fields, even if general knowledge is already in hand. Indeed, one potential concern is the operation of those staffers that lack the specialized knowledge necessary to act responsibly and deftly carry out duties of the job. Faulty purchasing or development decisions can be costly to governments of all sizes, but at the level of local government the results can be ruinous. A new employee in these areas may not know the difference between a situation where one closes the contract or where one should simply walk away from it. Through training, real-world understanding of these specializations can be made apparent without the costly mistakes that inform some experience.

Public procurement has been around as long as the notion of government, but its professionalization is a more recent circumstance. The Packard Commission suggested that quality public procurement would entail "short, unambiguous lines of communication among levels of management, small

staffs of highly competent and professional personnel . . . [and] most importantly, a stable environment of planning and funding."[67] The emphasis in
such a framework is on stability, accountability and competence. Economic
development, or at least its concern with unemployment and community
investment, is also nothing new, but like public procurement, efforts for
practitioners in the field to agree upon standards of conduct, a common
body of knowledge, and professionalization and continuing education in
the field are all relatively new. As a result, these fields are likely emergent
professions, as they do not respond favorably to all the requirements of the
sociological interpretation of a profession,[68] but are clearly affirmative in
important aspects including the basic points outlined by Brandeis.

There is skill involved in both public procurement and economic development, but as governments find out more about both areas, there is an
increasing realization that knowledge is a key part of success. Both fields
succeed on Brandeis's first point. On his second point, about work being
done for others, the case can be made that the public interest is in mind in
government purchasing as it is in economic development. This work, where
the practitioner plays the role of advocate for the larger public good, is
done for others, and staffers must resist attempts from a variety of sources
to be swayed toward unethical or corrupt ends, given the power associated with the positions. On his final point, about the importance of financial compensation vis-à-vis the value of the job itself, while government
employees perform relatively well in pay and benefits, the primary reason
for engaging in these endeavors is probably not money alone. The values
statements for both National Institute of Governmental Purchasing (NIGP)
and International Economic Development Council (IEDC) suggest that
working in service of the public interest is a core principle. Of course, some
practitioners in both fields will run afoul of the high-minded ideal of public
service in seeking personal gain, but there is professionalism in purchasing
and development activities, as there is in other forms of administration in
public service. For those that would engage in corrupt activities, it is possible that no code of conduct would deter them anyway.

The problem with public procurement and economic development is that
these have not been seen by the public historically as professions in the
traditional sense of the term, as would medicine or law. The difficulty is
compounded by the lack of information the public has about what constitutes the work of the two groups. Public procurement might bring to
mind the $800 toilet seat purchases that at one point made governmental
purchasing the subject of jokes and gave rise to the sense that purchasing
in government amounted to fleecing of the public. Economic development
has recently had its share of negative publicity, including a series of articles
in *The New York Times* about the excesses of incentives programs. The
series intimated that local governments forego $80.4 billion in revenue each
year because of incentives programs.[69] It is not unreasonable to suggest that
the general public does not understand economic development, and because

there is a lack of understanding, spending in this area is seen as suspicious and further evidence of government run amok.

The value of professionalization in public procurement has been illustrated.[70] Public procurement has a positive role to play in disasters, and communities benefit from a strong local purchasing group which demonstrates a sense of openness and responsibility shared by the institution as a whole.[71] Failing to provide for professionalism in the purchasing function within local government is a mistake that may lead to difficulties in extraordinary situations, and will probably hamper the ability of local government to respond effectively. The reason for this is that the purchasing group can work closely with local businesses. If the purchasing group understands the needs of local businesses in disaster, it can be a part of getting them up and running in response and recovery. The purchasing group can also anticipate community needs and be more responsive to them; the role of local government in providing a direct connection to the community, where citizens and local businesses can see the response of local government and understand that connection, is essential.

The value of professionalism in economic development is equally important. Economic development professionals, with their understanding of the business community, can play the role of experts for local government in identifying needs in the business community and assisting leaders in defining a vision for a community's future. Entrepreneurship and the growth of local businesses, planning for continuity of operation in the event of a hazard event, and the support of businesses in crises through business recovery initiatives are all excellent initiatives that can be provided by economic development professionals. Both procurement and economic development require professionalism, not only for legitimacy of their roles, but because of the potential for corruption and the need for responsible and accountable behavior in these specialties. Both fields have far-reaching implications, for government, business communities, and individuals within the community, through interactions with government and businesses as citizens and employees.

In the next three chapters, we examine cases of communities responding to hazard events. The three places are very different from one another in practically every respect. The unifying theme is that local government has a role to play, positive or negative, in the future of communities confronted by hazard events, through attention to community resilience.

# 5　The New Orleans Region and Hurricane Katrina

*Most of the area will be uninhabitable for weeks . . . perhaps longer . . . at least one half of well constructed homes will have roof and wall failure. The majority of industrial buildings will become non functional . . . Airborne debris will be widespread . . . persons . . . pets . . . and livestock exposed to the winds will face certain death if struck . . . Power outages will last for weeks . . . as most power poles will be down and transformers destroyed. Water shortages will make human suffering incredible by modern standards.*[1]

Katrina hammered the Gulf Coast of the United States with abandon on August 29, 2005. The storm had weakened significantly from its previous Category 5 fury; at first, New Orleans thought it would escape relatively unscathed, but this was not to be. As the ominous National Weather Service bulletin warned, damage was astonishing, but the suffering experienced in New Orleans went well beyond the toll exacted from the community—its individuals, businesses, and governments—by the storm itself.

Study of Hurricane Katrina suffers in that there is little objectivity in analysis. Perceptions about what happened are as close to reality as some are willing to come in describing what occurred, particularly in New Orleans, and why it happened at all. The discussion is laced with pain; many are unwilling to discuss the matter further at this point, leaving us with *petits récits*—small narratives with limited validity when generalized beyond an instance. This lack of objectivity impacts research efforts in unfavorable ways. Community recovery, and specifically the response of businesses, after disaster is an important part of disaster research, and this aspect of understanding hazard events deserves greater attention. We cannot get to a full understanding of the event, and potential applications of knowledge to situations like it, without objectivity.

In this chapter, the focus is Hurricane Katrina and the experience of the New Orleans Metropolitan Area, including the parishes of Jefferson, Orleans, Plaquemines, St. Bernard, St. Tammany, St. Charles, and St. John the Baptist. The chapter begins with an overview of the Katrina event, and provides background for the City of New Orleans and the region, and how hurricane response was managed by local government. An examination of the results of a survey of business owners that had registered as interested in government contracting work follows. Finally, the perspectives of public officials on issues of vulnerability and resilience in the City of New Orleans and the surrounding region are brought to the forefront. The results of the mixed methods approach suggest an area with high vulnerability before and after the storm and low resilience that is systemic.

## HURRICANE KATRINA

The year 2005 was a historic season for tropical weather in the Atlantic Ocean basin. It was the most active hurricane season on record, with twenty-eight tropical cyclones, and seven major hurricanes, including four Category 5 storms—Emily, Katrina, Rita, and Wilma.[2] Three of the six most powerful hurricanes to that point in 153 years of modern weather history had occurred within a period of just fifty-two days.[3] Katrina, the worst storm in a season of terrible tropical events, impacted about 90,000 square miles and displaced 400,000 people.[4] The storm's 140 mph winds breached "nearly every levee in metro New Orleans," where the flood protection system failed in fifty-three separate locations.[5] It is not possible to underestimate the tragedy, in societal, economic, and human costs, even years later. The sociocultural disruptions to communities and lives in the affected region were extensive and even complete

*Figure 5.1*  Flooded neighborhoods in New Orleans, August 30, 2005, FEMA/ Jocelyn Augustino.

in some cases. The images of flooding in the area, as shown in Figure 5.1, are particularly memorable.

Katrina is responsible for the deaths of 1,833 people, "1,577 fatalities in Louisiana, 238 in Mississippi, 14 in Florida, 2 in Georgia, and 2 in Alabama."[6] Approximately 35,000 people had to be rescued due to the storm's impacts.[7] The uncertain track of the storm in the two to three days before the landfall led some to a false sense of security in the face of the pending destruction, and a relatively late evacuation, leaving about 100,000 people stranded in metropolitan New Orleans as late as 4 PM on August 28, 2005, the day before the storm arrived.[8]

The disaster affected communities in Louisiana and Mississippi, where the unemployment rate doubled from 6 percent to 12 percent and salaries fell by a combined $1.2 billion.[9] Between 1.6 and 1.7 million people resided in Katrina-affected areas in Louisiana and Mississippi.[10] About 1.1 million were evacuated due to the storm, and 770,000 people were displaced by its effects.[11] By late December 2005, 500,000 of those people still had not returned home. Among evacuees not returning home, the unemployment rate reached 28 percent.[12] Year 2008 estimates were that New Orleans would only have 41 percent of the jobs it had before the storm, and even if jobs existed, 150,000 to 160,000 homes were damaged beyond repair or completely destroyed.[13]

According to a 2011 estimate, Katrina caused $108 billion in damage, making the storm "far and away the costliest hurricane in United States history."[14] Adjusted for constant 2010 dollars, the damage is $105.84 billion.[15] Staggering estimates of the total economic impact range from $200 billion[16] to $300 billion.[17] Even if counting an exact cost is difficult, the immensity of the event and the level of the suffering it caused are persuasive enough.

The role of people in exacerbating the situation post-Katrina in New Orleans pushes the issue far beyond that of a natural disaster or the impacts of a hazard event. What occurred in New Orleans, as has been widely noted, was not natural. Bungling mismanagement by government at all levels led to a crisis situation far worse than what should have occurred—the true disaster was essentially caused by people.[18] The literature holds all levels of government responsible for failure to protect the public. Local government appeared inept and incapable of handling any aspect of the storm's fury or its aftermath; the City of New Orleans in particular was ill-equipped to handle the onslaught. Faced with the devastation all around them (Figure 5.2 shows an example of the severity of the damage), city personnel fought to regain control but did not communicate a sense of competence to the public despite their best efforts. But the problems were not limited to any one level of government; there was no shortage of blame to go around. The response in the aftermath of Katrina goes beyond simple negligence or failure to act; some have suggested that abuses in the aftermath of the storm amount to state crime, because there was awareness of the threat and a lack of planning and action when the threat materialized as Katrina.[19]

*Figure 5.2*   Homes in the Lower 9th Ward of New Orleans were heavily damaged, FEMA/Andrea Booher.

What are we to make of Katrina and what happened in New Orleans, and how it impacted the business community specifically? What role was played by governmental institutions and culture?

## NEW ORLEANS: UNIQUENESS AND VULNERABILITY

To provide context for this study, one must understand the community and its businesses. According to 2012 estimates, New Orleans was a city of 369,250 that was 60.1 percent African-American, down from 67.3 percent in the year 2000, when the population was 484,674. The population increased 7.4 percent since 2010. An estimated 25.7 percent of individuals lived below the poverty line, which is a drop from 27.9 percent, the rate in 2000.[20] The unemployment rate among the African-American community is historically high—it was 13.1 percent in 2000, while the unemployment rate for the Caucasian community was 3.6 percent.[21] White families earned over twice as much as what African-American families earned.[22]

The jobs available in New Orleans have changed dramatically since 1964. Manufacturing jobs, which paid well, disappeared, and were replaced by relatively low-paying, low benefit service-sector jobs, such as those supporting the restaurant and tourism trade.[23] This change led to declining population in the city overall; with the decline of manufacturing, the white population of New Orleans began to leave. This scenario "helped establish and perpetuate the disadvantage of the majority of the city's residents."[24]

A 1987 article indicates that a city official had the opinion that "there's nothing wrong with tourism as a [development] strategy," and this appears to be the strategy that has been employed in New Orleans up to the present day.[25] This is despite the fact that there appeared to be, throughout the city's recent history, understanding of the importance of economic development and support of a variety of industries to ensure adequate, well-paying jobs. There were attempts to rejuvenate economic development, even in the years immediately prior to Katrina, with the creation of Greater New Orleans, Inc. The ten-parish regional economic partnership has dedicated itself to fulfilling the region's "potential as one of the best places in the country to grow a company, and raise a family. The ultimate indication of success for GNO, Inc. will be the existence of a robust and growing middle class in southeastern Louisiana."[26]

At the time of my research, only small incremental gains in diversification of economic development had been made. The city was still too reliant upon the tourism trade, and had little robust retail to speak of, beyond small shops and an upscale mall bordering the French Quarter. Its citizens had to leave the parish to do basic shopping for retail and groceries. There were no hospitals in Orleans Parish, so health care was provided in surrounding parishes for the most part; nevertheless, health care was a major employer for those who lived in Orleans Parish, so they had to leave the parish to go to work. This also presented problems when patients had to travel farther to get health care, especially in emergencies. Orleans Parish had significant leakage, to use the economic development term, in its own tax base—citizens spent their money elsewhere and the city had a great reliance on tourist dollars for funding. A logical question is the readiness of a city that is missing fundamental services within its borders, such as adequate health care, schools, and even grocery stores.[27]

Given that many residents in the city did not have ready access to transportation before Katrina, not being able to provide for basic needs can force already vulnerable residents into a spiral. Supportive groups that have gone into areas like the Lower Ninth Ward have found that government has still not met community needs with respect to livability or sustainability. One particular group, Global Green USA, noted that "residents . . . have to travel more than 3 miles into St. Bernard Parish to shop for fresh fruits and vegetables at the Wal-Mart in Chalmette."[28] People may want to move back into the area but the area is not ready to support them; individuals show signs of resilience, but in this respect it could be said government has not.

Other industries for the city's population were the source of some long-term concern, as well. The city continues to be reliant on its seaport and the petrochemical industry, representing foundational sectors of international trade and energy.[29] The 2010 Deepwater Horizon oil spill in the Gulf of Mexico is clear indication just how difficult matters can be when there is a disaster and the connection between an industry and an area is close. Interviewees were, during my visits to the area, concerned about those impacts—to the fishing community, primarily. In efforts to protect these other industries, such as fishing and provision of gulf seafood, local leaders have since made efforts to diversify the ability for these businesses to make money. What was evident from the experience with the oil spill is that local leaders were aware that the area is too reliant on too few sources of income. This economic concentration in few areas is indicative of catastrophic potential.[30] To continue in this manner does not provide for a sustainable future for residents of these areas.

For Orleans Parish, continuing to rely principally upon tourism does not allow for the citywide rebirth local leaders desire. Tourism is a hallmark of the city and the fun of New Orleans is its calling card, but without jobs that pay well and provide good benefits, the city has resigned itself to a future limited by its past. Urban studies scholar Richard Florida has commented that moving away from a tourism-based market is a key way of steadying an economy and avoiding boom and bust cycles.[31] Moving forward requires partnership with the business community, retention of existing businesses and attraction of new industry that has long-term potential, and a resolve to continue those things that are most attractive to what the city leaders are hoping to build. Another industry that has been a major economic engine around New Orleans for many years has been the petroleum industry, itself subject to boom and bust, even as it has made some segments of society quite wealthy. In New Orleans, there is a limited middle class, a great deal of poverty, and a tough life for many in the community. The city itself is a jewel, but perhaps has had its potential unfairly limited in the choices it has made.

One gets the sense that economic development efforts in New Orleans remained embryonic as late as 2010. In one interview conducted for this study, an official mentioned that the City of New Orleans cut economic development drastically after Katrina—"after the storm the economic development offices in our city were cut evidently from over 50 to somewhere in the teens and so they lacked capacity to institute all of the programs." The interviewee further indicated that, in addition to the city lacking capacity to institute economic development programs, it failed to work with the business community to enact programs that would work. New programs, it was indicated, were intent on "ensuring that that emphasis included buy-in from the business sector and from the public sector so that we can work and go in the same direction." But without a direction for economic development over the first five years after Katrina, the city, from an economic perspective, was clearly not moving forward in the most advantageous way possible. To the extent that business came back, the business community has been a prime reason for this outcome, with local government's involvement being limited over the years immediately following the storm.

We must also consider the impact of investment in an area when the risk is always present for additional hazards. Governmental efforts may be seen as attempts to provide for additional resilience, but the fact remains that New Orleans is in a perilous position—essentially a bowl of land surrounded by levees—and the question is not if a hazard event will befall the city again, but when. Katrina itself was not a catastrophic storm event, and it is worth remembering that fact; had the storm arrived as a stronger event, and had the city taken a direct hit, the result would have been even worse, inconceivable as that is. Even with its many charms and easy-going attitude, the city has a decidedly defensive posture for obvious reasons. It is an open question whether the area is sustainable, with subsidence and sea-level rise being important concerns. Consideration must be given to how coastal population centers are impacted by such issues, and Katrina provides evidence of the salience of such matters.[32] Further, environmental sustainability with impacts from the petrochemical industry remains an issue. One of the reasons long-run investments have failed to come to New Orleans may be that the city is in a risky position, and corporate interests cannot tolerate taking on such risks. Short-term propositions are an easier sell—the tourism aspect of the city is undeniable and events and conventions are a natural for economic gains.

The position of the enterprise of New Orleans, on the whole, is far from stable, and the message on economic development and investment is mixed. Where local government might have played a role in supporting business, at least in New Orleans itself, it is not clear local government has done enough or placed appropriate resources to assure a bright future for the city. Other parishes have fared somewhat better, given a keen business acumen and a willingness have an open and ongoing conversation with the business community that both reduces uncertainty and gains trust in the partnership between private interests and the public sector. This is not to say that New Orleans cannot attain such a position in leading the discussion; however, before, during, and after Katrina, and for a number of years into the recovery, New Orleans had not played this indispensable role. In its defense, though, it has had plenty of problems with which to contend.

A major area of vulnerability for the city is crime. New Orleans had a crime problem before Katrina and it had one after Katrina. The burglary rate before Katrina was 48.9 burglaries per 100,000 residents in the month before the storm. In the month after, the rate rose dramatically to 245.9—a four-fold increase.[33] New Orleans was and is a violent city; it has long been at or near the top of the list for murder capital of America, with homicide rates at ten times the national average.[34] In 2013, New Orleans made national news for a shooting that occurred on Mother's Day; shots were fired into a crowd of participants in a second-line parade, and nineteen people were injured.[35] Earlier in the year, shootings occurred on Martin Luther King Jr.'s Birthday and during Mardi Gras. Mayor Mitch Landrieu was quoted as saying, "Unfortunately . . . the specialness of [Mother's Day] doesn't seem to interrupt the relentless drumbeat of violence that I've talked about so much on the streets of New Orleans . . . It's a shame and it's got to stop."[36]

This violence has implications for economic development because business development accompanies, according to some theorists, the presence of a creative class. "Economic developers . . . have focused on attracting cool people—meaning healthy, active, educated people with money to spend—because they add diversity, energy, creativity and passion to a community."[37] The people New Orleans would like to attract generally do not wish to locate in places that have high crime and a lack of well-paying jobs. Further, the city has had much difficulty in gaining control of its gang problem, stemming from the robberies, shootings, murders, and associated mayhem caused by thirty-nine targeted groups.[38] This crime and the fear experienced in the general population from the city's inability to gain the upper hand create a sense of resignation among residents about the city's capabilities and its future, which becomes something of a self-fulfilling prophecy.

While crime is a problem, gentrification is also a genuine concern that threatens the character and soul of the city. New Orleans has lost much low-income housing due to Katrina and still has a high population of individuals below poverty to serve. Building gentrified housing on former low-income sites cannot be the answer, especially if the one response of the city is to remain a tourist destination as its primary economic driver. The people who work in the city in these tourist jobs cannot afford to call the city home. The cost of housing continues to rise and this unduly impacts already vulnerable populations. As noted by Barnshaw and Trainor, unchecked inequality prior to a disaster event results in a post-disaster reality where the inequality is greater than the pre-disaster period. More advantaged groups may reattain full capacity, but vulnerable populations may never regain their full pre-storm capacity.[39] This stratification is corrosive in a society and it exists in New Orleans. Any economic development program that does not consider at-risk populations, their skills and potential, does not fully enhance the prospects of the whole community and therefore runs the risk of further gentrifying it.

The greatness of New Orleans has always been its cultural heritage and diversity. New Orleans is at particular risk in maintaining this because of the breaks that occurred in the post-Katrina period. Whole communities, and especially vulnerable communities, were disrupted. Meanwhile, some of the communities that did best in terms of their recovery and regaining capacity were communities that relied least on the assistance of government. The Vietnamese community in New Orleans, representing about 23 percent of foreign-born citizens in the city, was able to regain considerable capacity and did so rapidly.[40] Examples of communities within the larger context of New Orleans exist where people "took care of their own," like the Vietnamese, but the threads obviously frayed in the aftermath of the event, and without that shared tradition, the more common response was that people were left to fend for themselves. This is consistent with Alesch, Arendt, and Holly's observation that "communities most likely to recover see themselves as self-organizing rather than reliant on an external agency."[41]

If vulnerable communities are not actively protected, as a central consideration and expected outcome, then gentrification may occur because of a focus on more purely economic outcomes—the attraction of different populations that may provide for a different skill set, for example,[42] or the belief that a city should create a different future for itself that removes vulnerable populations from the equation, rather than making those populations less vulnerable. These new populations may replace existing at-risk populations. In the New Orleans case, the city of tomorrow may be fundamentally different, with important cultural groups being crowded out by new groups that can afford the supposedly enhanced lifestyle of the city. The city has already had shifting populations over the years, with the formation of "superghettos,"[43] and other characteristics that are less than amenable to a modern, welcoming place. A government that governs best seeks to protect all its citizens and should not advantage certain of them for questionable ends. But it is apparent from coverage of both pre- and post-disaster redevelopment that not all groups within the city have a seat at the table for discussions of governance that have a community-wide impact. This does not create an environment that fosters community resilience.

All of this has direct impact on resilience among businesses and the community. Small businesses represent the vast majority of businesses in most communities and this is true of New Orleans and the surrounding MSA (Metropolitan Statistical Area). Small businesses are a source of innovation in the current economy and they are also the source of many new jobs. Communities that have governments and infrastructure systems that are supportive of business operations tend to retain and attract businesses more effectively. In a place like New Orleans, with a large minority population, inclusion is even more important, as technical assistance programs can be the difference between success and failure for minority-owned businesses. Vulnerability becomes a critical point of concern because it can determine a large measure of the ability of the community as a whole to rebound when challenged in disaster. As we will see, vulnerability was, and remains, a serious concern in New Orleans, among individuals and businesses.

## COUNTING THE PERSPECTIVE OF THE BUSINESS COMMUNITY

This project includes both quantitative and qualitative methods in analyzing each of the case areas. We begin with the quantitative approach, which explores the views of businesses through a survey of their perceptions of local government with an emphasis on the experience of firms in disaster. Business owner perceptions are also examined through a review of their responses to a survey question asking for general feedback.

The purpose of the survey of business owners conducted for this study is, as with other survey methods, to "generalize from a sample to a population so that inferences can be made about some characteristic, attitude, or behavior of this population."[44] Surveys were selected as a means of contacting registered vendor communities and asking questions about attitudes about and experiences with public procurement, because survey methods, particularly Web survey methods, have potential to contact large pools of potential survey respondents in a short period of time, at minimal cost. The common experience of vendors in responding to public procurement, and dealing with disaster recovery, utilized a set of terms that lend themselves well to questions that are widely understood among business owners. The philosophical framework for the surveys is pragmatic—the questions lend themselves to readily grasped concepts, like "which parts of the program work well and which need improvement? How effective is the program with respect to . . . beneficiaries' needs?"[45]

The survey instrument used was a self-administered, Web-based questionnaire, offered via email recruitment to vendors registered with the City of New Orleans, the Central Contractor Registration system of the federal government, the Louisiana state contracting list, or the Louis Armstrong Airport Vendor Database. For the non-city databases, vendors contacted for the survey were limited to those in the New Orleans MSA. There is a general expectation among vendors engaged in government contracting that e-procurement is increasingly the way of government purchasing, and even in non-e-procurement environments, contact via email for solicitations is common. In other words, using a Web survey should not unduly bias the result in measuring the views of vendors registered with a local government procurement system. Though email addresses and a voice mail number were included for those participants that had questions, questions were very rare. When I heard back from vendors, it was generally in the form of comments from the potential survey takers, ensuring that the survey was a legitimate academic exercise, asking questions about how the information would be used, expressing anger at having received an invitation, or, interestingly, offering unsolicited personal comment on interactions with government purchasing operations or requesting assistance on particular public procurement issues with given entities. I provide an example of the survey instrument used for the Palm Beach County case as Appendix A. The New Orleans MSA case used a very similar survey instrument.

The original intent of the project was to utilize the vendor database from the City of New Orleans and the Orleans Parish School Board, exclusive of one another for comparison's sake. Vendor directories were requested from the City of New Orleans, along with information on contracts awarded and projects undertaken by the parish and the parish school board. The responses from the City of New Orleans to these requests were incomplete. Requests for information went unanswered for a period of months, despite repeated follow-up contacts to various parties throughout city government. The vendor database eventually provided was an Excel spreadsheet, which lacked basic information for contacting firms electronically, such as email addresses. For the Orleans Parish School Board, no vendor list was ever provided, nor was contracting information provided despite repeated requests. Even inquiry to the Louisiana State Attorney General with a request to intervene did not result in a successful collection of contract information from either entity.

Because the surveying of firms in the New Orleans area required electronic contact information, I decided to widen the focus to the larger New Orleans MSA for purposes of obtaining survey information, and to expand the focus of the contracting questions in the survey to the other public entities that may offer contract opportunities. The vendor database information provided by the City of New Orleans was supplemented with local vendor information (firms located within the New Orleans MSA) for firms registered as vendors with the State of Louisiana and with the Federal Government Central Contractor Registration system. The compiled list, while not an optimal representation, nevertheless sought to gather as much information as possible on vendors that were interested in working with local government in the New Orleans MSA, while maintaining a focus on the City of New Orleans.

Selection for the survey involved all firms with valid email addresses being contacted to request their participation in the project. Each firm was provided a unique link that allowed them to complete the survey once. Because all firms were contacted, each firm had the same chance of being included in the sample. The responses were such that there is worth in each, particularly from the perspective of the disaster response data and business interaction with local government. What would be more important, considering the breakdown of vendors in a vendor registration database, is that the relative proportion of firms by size (total gross receipts) should match market-wide indicators for the local area; comparative data from the U.S. Census County Business Patterns database, which also utilized the general North American Industrial Classification System (NAICS) codes, would serve as a useful check on representativeness. As a general rule, most firms filling out the survey should be small, with firm size increasing as response rates decrease for each range of gross receipts.

I used the ZipSurvey.com online platform to send out the surveys. I sent 4,491 invitations to complete the survey via email; 342 invitations received a "click-through," where the recipient clicked the link associated with the invitation; and 186 surveys were completed. Consistent with the standards of the Interactive Marketing Research Organization, response rate is "based on the people who have accepted the invitation to the survey and started to complete the survey. Even if they are disqualified during screening, the attempt qualifies as a response."[46] We have no way of knowing how many of the 4,491 businesses even received the survey email, nor do we know how many did not receive it because of misdirected email, spam systems, or other considerations. We do know that 342 business owners/primary contacts opened the recruitment email, and having read it, 186 clicked through to the survey form online and entered data into it. This is a response rate of about 54 percent.

The reasons for the response rate not being higher include the use of Internet surveying and the decline overall in responses for this medium; lack of interest among residents in answering further questions about Hurricane Katrina; and the presence of closed businesses or faulty email addresses in the vendor list. Nevertheless, the lack of data on the business community's experience following Hurricane Katrina, and specifically its experience with local government procurement systems, encourages our review and analysis of the responses received.

Firms responding to the survey were given an opportunity to report with which parishes/agencies they had pursued contract opportunities. Respondents' indications can be seen in Table 5.1 (firms could choose more than one parish/agency):

Demographically, 67 percent of the firm owners responding to the survey were white; 18 percent were African-American, and 4 percent were Hispanic. Fifty-seven percent of firms were owned by males; 23 percent were owned by females, and the balance were held by co-ownership or other arrangement.

When asked about contracting practices with the parishes/agencies that they had worked with, business owners were more negative than positive in their perceptions. Assessing agreement with the statement, "When a firm is awarded a contract with local governments, it is based more on what they know, than who they know," 55 percent of firms indicated that they disagreed with the statement (35, or 28%, indicated that they disagreed, and 34, or 27%, indicated that they strongly disagreed). Twenty-nine firms (23%) agreed with the statement, "I am treated fairly on contracts with local governments, by local government staff." In response to the statement, "The actions of lobbyists

*Table 5.1*   Firms seeking contracts at New Orleans MSA parishes and authorities

| Parish/Agency | Firms Seeking Contracts | % |
| --- | --- | --- |
| Orleans | 121 | 20% |
| Jefferson | 95 | 16% |
| Plaquemines | 34 | 6% |
| St. Bernard | 46 | 8% |
| St. Tammany | 55 | 9% |
| St. Charles | 31 | 5% |
| St. John the Baptist | 30 | 5% |
| Orleans Parish School Board | 40 | 7% |
| Louisiana Recovery School District | 39 | 7% |
| Sewerage & Water Board | 38 | 6% |
| New Orleans International Airport | 38 | 6% |

are a decisive factor in the award of contracts with local governments," only nineteen firms disagreed or strongly disagreed (15%). This is consistent with other sources, which contend that patronage is a problem in New Orleans contracting, and that "many citizens of New Orleans have lost faith in the government's ability to procure services that are cost-effective and serve the public interest."[47] In a separate report, the Bureau for Government Research observed that neighboring Jefferson Parish's approach to services contracting had too much involvement by council members, leading them to find it an "idiosyncratic process that some perceive to be driven more by personal political relationships than by what makes the most sense for taxpayers."[48]

Speaking specifically about procurement in the aftermath of disaster, firms responded as follows to the following three separate statements, capturing elements of local procurement quality in that context.

1. I had access to government contracting opportunities that were related to hurricane disaster recovery.
2. For disaster contracts, the local governments protected the interests of local businesses in securing disaster recovery work.
3. During the hurricane recovery process, the award of recovery contracts helped to sustain businesses in my community.

Responses are shown in Table 5.2.

As a group, the responses were negative as far as overall impression of the procurement process and how it contributed to recovery from the perspective of business owners that responded to the survey.

The survey form included one free-form question, asking for general feedback or information not already addressed in the survey questions. The question proved to be a welcome addition to the survey, in that it provoked a variety of honest comments from the participants. As a group and as might be expected given the responses to the Likert-scale-based statements, the participants were highly critical of local government procurement practices, so much so that analysis of the material becomes a matter of the degree to which the firm owners are jaded and believe that local procurement processes are unreasonable, difficult, or even corrupt.

I coded the free-form comments to generate themes and categories of response. The coding step yielded four major categories of response:

1) Firm owner has a positive or neutral view of local government procurement (6 instances)

    Example: "What we have gotten has helped us stay afloat."

2) Firm owner has a negative view of local government procurement processes, due to program problems, apathy, or slowness in how government works. (21 instances)

Table 5.2 New Orleans MSA, responses to questions centering on disaster procurement quality

| Question | Strongly Agree | % | Agree | % | Neither agree nor disagree | % | Disagree | % | Strongly Disagree | % | NA | % |
|----------|----------------|-----|-------|-----|----------------------------|-----|----------|-----|-------------------|-----|----|-----|
| 1 | 6 | 7% | 14 | 17% | 20 | 25% | 14 | 17% | 28 | 35% | 19 | 23% |
| 2 | 5 | 6% | 2 | 2% | 23 | 28% | 21 | 26% | 33 | 41% | 17 | 21% |
| 3 | 2 | 3% | 10 | 12% | 28 | 34% | 13 | 16% | 30 | 37% | 17 | 21% |

Example: "The employees of [parish] have been very polite and easy to work with. The process has been very slow though."

3) Firm owner suggests that it is "more who you know than what you know" in getting local government contracts. (14 instances)

Example: "Local governments procure based upon who you know."

4) Firm owner believes there are major problems in public procurement, cannot keep their business afloat, or openly alleges corruption. (27 instances)

Example: "Most of the prime contractors will use our bidding numbers or our minority status to get the contracts but will not hire our firms to do the work." "Local government was totally incapable." "St. Tammany was the best in the way it awarded contracts and Orleans was the absolute worst. It was like dealing with a third world country."

5) Firm owner will not pursue local government contracts due to the negative connotations of the purchasing environment. (3 instances).

Example: "Our firm still won't touch state/local business with New Orleans/Louisiana."

Where firm owners wished to speak on the subject of local procurement, comments were generally not positive. There is general agreement among respondents that there were considerable problems in local government procurement, with New Orleans being singled out of the group by more than one respondent. The bitterness and resentment was apparent.

We now turn to the results of regression analyses for business recovery.[49] Problem severity was quantified as size of disaster in two ways: damage to business, and whether the individual business owner noted that personal damage impacted business recovery. The business damage response was transformed with a square root transformation to normalize the responses, which had a large range. The personal damage response was coded as a yes or no response. Sixty-six of the 186 firms in the survey reported some kind of response on business damage, even if the response was that they had no damage to report.

Disaster Procurement, Small Business, and Institutional Culture Factors are each based upon a factor analysis of questions that with responses exhibiting collinearity within the thematic material of each factor. Table 5.3 details the factors and questions they represent:

For the vulnerability measures, minority and gender as applied to ownership of businesses are indicator variables, as firms are or are not owned by minorities or women, and coded as such by their responses. A firm's response

*Table 5.3*  Institutional and institutional culture components from factor analysis

| Institutional Factors | Component Questions |
| --- | --- |
| Disaster Procurement | I had access to government contracting activities that were related to hurricane disaster recovery. |
| | For disaster contracts, the county/parish protected the interests of local businesses in securing disaster recovery work |
| | During the hurricane recovery process, the award of recovery contracts helped to sustain businesses in my community. |
| Small Business | Certification [as a small business] has helped my firm win prime contracts that it would not have won otherwise. |
| | Certification has helped my firm win subcontracts that it would not have won otherwise. |
| | Since I became certified, my work on government contracts has increased significantly. |
| Institutional Culture | When a firm is awarded a contract with this county/parish, it is based more on what they know, than who they know. |
| | My firm has a strong working relationship with elected officials in this county/parish. |

on whether it had insurance is also an indicator variable. Size of business was coded according to a scale of annual gross receipts size ranges, from a potential response of "less than $25,000" to "$100 million or more" in fourteen incremental ranges. The mean of 5.89 is placed in the $250,000 to $500,000 range for gross receipts on average among respondents, which is as expected given the prevalence of small businesses in the market compared with all firms.

Sample population statistics, including distribution of respondents by primary industry category; gross receipts of firms, by annual average, for 2002–2004; number of local government prime contracts received; and number of local government subcontracts received are included with this study for the surveys conducted. The sample population statistics for the New Orleans MSA, as well as correlation and regression tables for the OLS regression[50] utilized for this analysis, are available online.[51] Table 5.4 summarizes the quantitative findings of the regression analyses for New Orleans.

The analysis found that problem severity, or the size of the disaster, is significant. Both variables tapping this construct, business damage and personal damage, are statistically significant. Personal damages impacting business recovery has a higher level of statistical significance ($p<.01$). The expected direction of the effect is as we would expect—the impact of increasing extent of damage on a business is negative, and results in a longer amount of time to return to normal.

*Table 5.4* Summary of significant findings, New Orleans MSA, regression analyses

| | Measures of Business Resilience | |
| --- | --- | --- |
| Study Area | Months to Return to Normal | Aftermath Capacity |
| New Orleans MSA | Damage to Business*; Personal Damage affected Business Recovery**; Institutional Culture*; Minority Ownership of Firm***; Size of Business* | Institutional Culture*; Size of Business*** |

*p<.1, **p<.01, ***p<.001

Institutional culture is statistically significant in impacting the months it takes for a firm to return to normal. Interestingly, neither governmental institutional indicator (disaster procurement or small business) is statistically significant in this application of the model, for this dependent variable. This lends some credit to the idea that the business community felt more on its own after Katrina and that there was little connection between institutional action and business recovery, though institutional culture, in this case the pervasive negative perception of local involvement, may have played some role in constraining business recovery.

For measures of vulnerability, minority ownership of firms impacts months to normal as a resilience measure. We expect from the literature on disaster vulnerability that minority groups have a higher level of vulnerability and have reason to believe that this may impact recovery, and in this case, we find that to be true. The result is statistically significant at a relatively higher level ($p<.01$). While the model does not rely on politics, the issue of race in political circles has historically been important in New Orleans and the surrounding area. The survey questionnaire included questions of a more political nature, but explorations with early versions of the model did not find them particularly impactful upon the indicators of business resilience examined, when considered in combinations with more settled elements of vulnerability and disaster impact. The negative undertone of some elements of the qualitative analysis, discussed later, does speak to some issues of political importance.

Size of the firm, as quantified in gross receipts, for the period preceding the storm, 2002–2004, is also significant. However, we would expect that size of firm would be important in that large firms would be more resilient. Here, the result counters our assumption. The positive connection we expect is actually negative, meaning the larger firms do not necessarily have shorter times to return to normal. This may be because small firms, with fewer employees, may be quicker to mobilize in disaster aftermath. If their location is undamaged, they may open and effectively return to normal, where a large business might find disrupted supply chains, branch offices unavailable for months, employees unable to work, and a host of other concerns to field. It is not the expected result, but it is explainable.

The second variable measuring business resilience is the percentage of capacity of the firm in the aftermath of the disaster. For this second approach to business resilience, using aftermath capacity as the dependent, only institutional culture and size of business are statistically significant. Size of business, for this dependent variable, is statistically significant at a high level. A change of note from the analysis for the dependent "months to return to normal" is the change from the institutional culture factor to a single component question "When a firm is awarded a contract with this local government, it is based more on what they know, than who they know." While the factor did not trigger a statistically significant outcome in reviewing this particular model, this question alone, as a proxy for the component, did.

It does appear, for the case of the New Orleans MSA, that the culture of institutions impacted business resilience, but the institutions themselves, procurement functions and small business initiatives, did not. Small business programs did not appear to have an effect, and the impact of local governmental purchasing for disaster purposes did not appear to have had an impact on either measure of business resilience. The fact that these programs did not have an effect, when we expect that they should, in part necessitated the additional study through qualitative methods, which follows this section. Size of business impacted aftermath capacity at a statistically significant level, with larger businesses seeming to maintain higher aftermath capacity, as would be expected. Neither variable for size of disaster, business damage or personal damage affecting business recovery, appeared to impact aftermath capacity as a dependent variable at a statistically significant level. In terms of vulnerability, neither of the demographic indicators had any significant relationship to aftermath capacity in this model. The fact that the model was not finding significant responses where we might expect them is, in itself, important and worth noting.

Institutional culture is reflected as significant in both measures of resilience, but the measure had a net positive effect in one instance (aftermath capacity) and negative effect in the other (months to return to normal). This could indicate that the results are uncertain, or that we are seeing two very different responses and that both are meaningful. For aftermath capacity, the positive response in impact of institutional culture is different than the result for long-term recovery, defined as months to return to normal; we are measuring two different aspects of resilience in the community. Aftermath capacity in the response period may have been affected by a variety of variables, beyond those examined in this study, including community networks. A positive response for the institutional culture question may have something to do with the connection between those firms that showed some level of capacity in the aftermath that also believe in the fairness of procurement opportunities. The long-term impact of lack of resilience would show strongly among firms that had the opposite feelings toward openness in governmental operations, generally. There may be divergence in outcomes, and this might explain the difference.

There seemed to be other issues at play, especially with regard to institutional culture and how this might impact how and the extent to which businesses might recover after disaster. The missing level of detail resided in the institutions themselves, in the official actors that were responsible for the institutions and that were either being encouraged or constrained by the prevailing institutional culture.

## INTERVIEW RESULTS AND ANALYSIS

As previously mentioned, this study includes an analysis of interviews conducted with public officials in New Orleans and surrounding parishes. For the New Orleans Metropolitan Area, eight interviews were conducted with personnel or officials representing the City of New Orleans and/or the Orleans Parish School Board. Two interviews were conducted outside Orleans Parish but within the Metropolitan Statistical Area. The New Orleans area interviews were conducted in July and August, 2010.

Interviews were requested from a broad range of public officials by contacting administrators and officials directly, via postal mail and email and via follow-up by phone. Questions regarding the nature of the inquiry were addressed and confidentiality of the responses was assured. As a result, all interviewees in this case and throughout the book are referred to as "an official" from the particular case area in question, without further identifying language. Interviews in New Orleans were conducted in person and face-to-face with the interviewee, with one exception that was conducted by phone due to scheduling conflicts.

I found that gaining access to public officials in the area was difficult; it appeared to me that there was a mistrust of what I might ask and how the material might be used. Some officials simply refused to speak to me or refused through omission—they were constantly unavailable and did not return phone calls.

The interviews made use of a questionnaire, which was altered to suit the particular job functions of the interviewee but otherwise held basic themes, such as fairness and effectiveness of public policies and programs, constant.

The approach to the interviews is cross-sectional—each respondent was interviewed once. If a certain question provoked discussion or comment, I pursued the discussion to gain additional understanding from the interviewee. Each interview was unique and representative of the particular subject's perspective, their institution, the disaster event and its impact, and the operational context of all factors. Certain interviewees used the interview as an opportunity to discuss what might be described as institutional constraint bordering on hostility. Others described an environment where they were encouraged to note inconsistencies in processes and to engage in discussions that would ultimately better serve the public interest. See Appendix B for an example of interview questions.

Transcribed interviews from the cases became the project's field text reports.[52] The reports were examined for the open coding step,[53] using the MaxQDA program, using line by line review to generate categories noting points of interest. The categories were then related to categories that bridged the open codes, through a process of axial coding.[54] Finally, through selective coding, theory is developed that fits "the topic and disciplinary area studied," is "complex enough to account for a large portion . . . of the variation in the area studied," and is "understandable and useful."[55] Interviews were taken as a group and analyzed for themes and commonality that could support the construction of theory. As concepts emerged, from axial to selective coding, a system of codes developed that yielded a working thematic base.

We begin with a discussion of the overall themes derived from the interviews. Using grounded theory, the codes of the analysis centered on issues of vulnerability for businesses, and inclusion and transparency within official processes. Within these general concepts, discussions tended to fix on phenomena that increased business vulnerability, while decreasing inclusion and transparency of official processes, or experiences that suggested the opposite.

The loose questionnaire that formed the basis of the talks with public officials for the New Orleans MSA and Palm Beach County cases yielded four major themes: two involving vulnerability, and two involving transparency and accountability; these results are shown in Table 5.5.

As is evident from the top-level themes in vulnerability, the reduction of vulnerability, for both case areas, involved a well-trained staff, a wide variety of open contract opportunities that invited the participation of the community, a financially sound and capable government that was aware of the threat posed by hurricane events, and the presence of preparedness plans, including emergency contracts. Curiously, in the case of the New Orleans area, entities outside the City of New Orleans spoke in greater detail, and with much greater frequency, about the presence of positive indicators that, given our understanding of vulnerability in the literature,

*Table 5.5* Vulnerability themes

| Reducing Vulnerability | Increasing Vulnerability |
| --- | --- |
| Staff in place for a long time/well-trained | Institutional vulnerability |
| Many contract opportunities available | Socioeconomic vulnerability |
| Government is financially sound | Natural vulnerability |
| Good intergovernmental relations | |
| Government involvement/effectiveness | |
| Hurricane preparedness: lessons learned | |
| Entity has emergency contracts in place | |

would reduce the threat of a hurricane event through governmental capacity and readiness. The New Orleans interviews spoke to awareness of vulnerability on a number of levels, including institutional, social, and natural vulnerability. The variation between the thematic elements of discussions with interviews in the City of New Orleans, and those from surrounding parishes, is surprisingly large. As an exploratory study, the information presented does not form a large enough base for broad generalization or prescription on its own, but the variance between processes and prospects in Orleans Parish and in those of surrounding parishes was notable.

In the interviews, procurement systems were often framed as either evidence of a government being inclusive and transparent, or provided as evidence of a lack of accountability in instances where the system was thought by the interviewee to be out of touch with what was needed.

Apparent from Table 5.6 is the commonality between the positive aspects seen in themes from the interviews and traits of resilient communities. Denial of problems and inability of institutions to run programs effectively, on the other hand, are harbingers of technological disasters and may be indicative of sensemaking failures within the organizations, in addition to lack of strategic vision from the top of institutions. The deep differences in responses from one

*Table 5.6*  Transparency/Accountability themes

| Inclusion and Transparency | Lack of Transparency/Accountability |
| --- | --- |
| Effective Local Small Business Programs | Budget crunches make running procurement difficult |
| Optimism/vision of a close-knit community | Budget shortfalls favor ending small business programs |
| Public & private sector worked together in response to disaster | Favoritism/Corruption |
| Businesses coming back to the community/reopening | Business closure after disaster/businesses did not reopen |
| Many new firms involved in the procurement process | Few contract opportunities available/very competitive |
| Official understands and has relationships with community | Ineffective administration |
| Fair procurement systems | Concerns about procurement systems |
| Economic development programs | No local business program |
| Transparency & accountability as goals or results of government systems | Small business program ineffective |
| | Denial of problems Gentrification/ land use/home rule concerns Lack of employees in institution, cannot run programs effectively. |

case to the next—even from one interview to the next—are strongly indicative of deep-set differences, in culture and policy, between local governments.

Table 5.7 reflects aggregate counts of codes and resultant themes from all interviews, by unique area within the New Orleans MSA, from the interviews conducted there.

Although the interviewees included a variety of officials and administrators in different roles, the themes that were raised were similar in both groups. In effect, a continuum formed of vulnerability, inclusion, and transparency, across the cases. Taken alone, the interviews provided for harrowing stories of the tragedy of Katrina. The discussions with the interviewees provide depth and context for the quantitative results. We saw in the survey results that institutional culture impacted business resilience. From the qualitative results, institutional considerations are shown to be complex and concerning. First, there is a sharp divide between the City of New Orleans and the surrounding MSA in the results of the qualitative analysis. Second, the lack of inclusion and transparency in the interview responses from the City of New Orleans, coupled with the increasing vulnerability themes, is troubling.

As can be seen above, for the New Orleans MSA, the interviews as a whole tended more toward thematic points that spoke to issues of increasing vulnerability, and reduced transparency and accountability. This is not to say that the entire MSA exhibits such issues. It is equally clear that the two neighboring parish interviews focused more heavily on reducing vulnerability and providing for open, accessible, and transparent government. Those interviews included active demonstrations of systems that supported such efforts. The discussion in New Orleans, perhaps due to the trauma of the event and the institution's lack of capacity to deal with it effectively, showed evidence of not being able to calculate the toll of the past.

It should also be noted that the examples given in interviews in this region of corruption and lack of accountability were disquieting, frequent, and bordering on unbelievable. However, given the nature of this particular disaster, and what occurred in New Orleans, the stories told were credible. It became all the more apparent comparing interviews from within the region with those from Orleans Parish; the sense of the disconnect between Orleans and other parishes is more or less complete—as if there were an administrative wall around the city.

Setting aside for a moment the idea of what has happened in the past, with the many and varied cases of public corruption that have been seen in the area, it is worth noting that the people who are responsible for government in New Orleans now are very much aware, perhaps even fearfully aware, of the responsibility they face with regard to making government accountable in the city. They were open and honest in interviews. When asked about a shift toward accountability in New Orleans, one official commented, "the idea would be let's get in here and create laws that would show unmistakably that New Orleans does its business in the light of day and to encourage the agencies and forces in our government that have the duty to investigate fraud and abuse." Processes that tended toward a lack of accountability were recognized:

Table 5.7  Incidence of major themes from qualitative analysis, New Orleans MSA

| | Reducing Vulnerability | % | Increased Inclusion and Transparency | % | Increasing Vulnerability | % | Lack of Transparency/ Accountability | % | Total |
|---|---|---|---|---|---|---|---|---|---|
| New Orleans (City/School Board) (8 interviews) | 4 | 1.50% | 107 | 39.30% | 14 | 5.10% | 147 | 54.00% | 272 |
| New Orleans MSA (Outside Orleans Parish) (2 interviews) | 3 | 2.90% | 70 | 67.30% | 12 | 11.50% | 19 | 18.30% | 104 |
| New Orleans MSA (Total) | 7 | 1.90% | 177 | 47.10% | 26 | 6.90% | 166 | 44.10% | 376 |
| *Average of Themes per Interview* | 0.7 | 1.90% | 17.7 | 47.10% | 2.6 | 6.90% | 16.6 | 44.10% | 37.6 |

"[contracting is] always going to be somewhat political but at least it's not so close to the politicians, it's been removed one step away so that there's an opportunity to actually have a fair process." Regarding the existence of latent problems with city administration, one official commented, "there has been no accountability by city administration at all in my . . . years being with the city. You report the failure say of a director to ensure that the procedures, city charters have been followed, and when it's not followed the answer is 'Well, they've already done the work so we have to pay them. So we'll approve it.' You know, there's no accountability. Nobody's held responsible for what they did."

Officials wanted business to hear that they were serious about the changes that were being made: "I would tell the business community that . . . they can look at the laws that we're putting on the books right now . . . and they will see that we are putting teeth into enforcement of these ideas of transparency and these ideas of fair treatment . . . that they can look at the programs and initiatives that the mayor is announcing on a weekly basis and then they will know that this is more than just a new administration and a new council speaking. We're not just speaking, we're acting."

The small business program was criticized by several officials. One official commented that it was a "good law," but "the problem is . . . enforcement of the law, the past administration had one person in an office that was monitoring all of the city contracts . . . [for] adherence to the law. I mean that's not enough resources put to that topic." The program may have had no impact because of its enforcement and the resources allotted to it, and been relegated to a symbolic role or one of patronage. Others commented that the program was enacted under threat of litigation and that this negatively impacted how the program was designed.

The stress and strain of Katrina and what occurred subsequent to the storm was still apparent, though. Some of these officials still appeared shell-shocked and weary when confronted with questions about what happened, and palpably wanted to tell the story to ensure that the tragedy of Katrina never happens again. One commented, "our first thing, and it gives me nightmares thinking about it, coming back we had to first dewater the city, get assessments, and then after that we had to figure out what areas would open up first, then what are the needs—the central needs—your gas stations, grocery stores, pharmacies. So those are the ones we went after first. The things you would need for you to stay here."

Importantly, cuts in staff after the storm resulted in a heavy burden for those remaining: "We had a staff of . . . wow . . . over seventy and after Katrina, we came back with a staff of less than ten," one interviewee reported. There were several reports of "bare bones" staff impacting the ability to get work done. Those who work in trying to keep businesses in New Orleans rely on the charm of the city: "It's very difficult to retain, the lure of the city that helps you keep the business here and the promise of making money."

Even where there have been lapses of policy, the staff that were on the front line seemed to know best practices and understand what they should be, but point out, very deftly, how the institution constrained their behavior and refused to budge on issues like changing policy or making government more responsive generally. There were specific instances in the interviews, which I will not point out for confidentiality reasons, that show how much institutional culture created stasis and actively prevented proactive change and ownership in governmental process. Actors within institutions, particularly in New Orleans, were not improvisers and did not seem adaptable, likely because the institution had limited them over the long term.

What can be pointed out is that little had happened in economic development in the five years after Katrina. One official noted that New Orleans "lacked capacity to institute all of the programs that [were needed]—we currently have a lot of monitoring functions and to initiate retention programs on top of that—so because of the storm the city hasn't had a major focus on retention and it wasn't so active, as active as they would have liked to have been and that's a capacity issue and the changes were a result of recognizing that and trying to make up for it." The official went on to suggest "the city does not have the capacity to do it [economic development]" and that the work might be better done in partnership with business, "giving them sort of an ownership stake to ensure they are successful so that our efforts are aligned and that we're working together and in the same direction." Partnership with the business community is a highlight of effective economic development programs and a hallmark of resilient communities. The same official spoke of the possibility of increased development planning and even targeting industries for growth. These would constitute positive changes, if they fully came to pass. But they would be unusual occurrences in the course of history in New Orleans, where similar commentary has been heard over the years.[56]

The issues faced by the City of New Orleans in particular are old issues, and the history of the city and its politics need not be brought up here, as there are already capable treatments of it.[57] One official commented that "we still struggle with topics that are Katrina-related but many of which are also pre-Katrina. Our blight situation . . . we have over 50,000 properties in a blighted condition that existed in many ways before Katrina. We're struggling still with a crime problem in the city that we're trying to change both on a short and a long term basis." So while it understood that it faced a tough climb, the problem may be that the city did not appear to be making the tough choices necessary to deal with blight and address crime, while visioning a future for itself. If the city failed in some ways, and it clearly has, it has failed in that respect. Katrina merely pointed out the failure, of the need for enlightened and responsive government, and for a fully wrought economic development program that can provide for quality jobs to create a better sense of place and a stronger community. It was not just that people needed government to save them, in Katrina; people needed government to govern.

Within city government, there were and are many people willing to admit that things can be done better—and should have been done better. It is not the people, perhaps, that were most at fault, because they put everything they had into serving the public, and I have seen that firsthand. It may be the institutional culture as the survey results suggest, and if the interviews are to be believed, that kept people from making sense of their situation and moving forward. Personal ethics carry the day most of the time in city government; some staff were obviously serious about changing how business is done, but there can be no doubt that others were more invested in keeping the system as it had been. Katrina provided more than enough evidence of this, within government and on the streets with people trying to survive.

## SYNTHESIS AND DISCUSSION

Katrina posed a variety of fundamental questions for community resilience that are worth reviewing. The purely organic response of local businesses has been inadequate to the challenges faced by communities in the storm's wake. As of this writing, New Orleans is trying to answer the same economic development question that has confronted it for many decades, and developing an answer for its future that is something other than tourism and energy has been tricky. Small businesses as a group, in the communities affected by Katrina, did not know what was necessary to adequately address and plan for the needs and trials of post-event reality. If we are shocked by any aspect of how businesses were impacted in Katrina, the most reasonable point for disbelief might be that the devastation was not much worse. It is reasonable to conclude that local government did not help as it could have, and may have made matters worse, for both individuals and businesses. The failure at other levels of government to provide assistance in a timely manner does not negate the failures at the local level. Surrounding parishes did better, but even they did not respond in a manner that could be fully considered resilient.

In the years since Katrina, the parishes have pursued approaches, to varying degrees, to try to make sure that failures of the type experienced during and after Katrina never happen again. The officials I spoke to seemed like good people who were genuinely interested in improving matters for their parishes and agencies, and while that was inspiring, the disconnect between what they want to do and what institutions will allow them to do is apparent. A lack of trust among levels of local government seemed endemic, and this inability to rely on a broad range of officials when confronted with threat or trauma is confounded by institutional constraint on individuals. This does not bode well for future threats from hazards, natural or otherwise. New Orleans has yet to fully overcome the miasma that muddled its response in 2005.

It has been suggested that small business owners in the Gulf area affected by Katrina engaged in minimizing behavior; while they heard that the storm was coming, and that it was going to be severe, they still failed to recognize the extent of the potential damage that they would endure.[58] Unrealistic optimism is dangerous,[59] and representative of a place where official actors cannot make sense of their contexts. From a business perspective, the result was a "preventable crisis" that could have been avoided—business owners could have moved more valuable merchandise out of harm's way, backed up records, or taken other steps to assure a quicker recovery after the storm. Firm owners were unable to assess damage quickly and their efforts were hampered by out-of-area people entering the region to assist with clean-up, who in turn congested roadways already strung with debris and used resources already in short supply, such as gasoline. Katrina was a "high consequence event"[60] and small business owners had to consider carefully how their response actions would affect their intermingled business and personal interests.

In disaster, small business owners are frequently in the predicament of not fully understanding what they are facing. There may be no other events of comparable magnitude that have affected an area for many decades. Small business owners trying to recover are forced to make quick decisions with long-term implications; because these decisions were being made with incomplete information, even the most rational approach is bounded by what is not known. For these choices under duress, error in decision is potentially compounded until it is not possible to chart a path to recovery. Unlike large businesses, with operations spread over regions, nations, or even internationally, small businesses may not have developed resourcefulness and adaptability, or the ability to improvise. Communication linkages in the community, with government and other businesses, may not be strong; this can hamper recovery efforts.

Even the decision to reopen is made more difficult by circumstances of a disaster. It may seem as though a business would need to reopen immediately after a hazard event, but the close relationship of small business finances to personal finances must be considered along with the potential for business success in the new post-disaster environment. A business owner might also consider how his or her business model had been impacted by the hazard event, and how to get to a "new normal" of business operation. The best decision for a small business after a catastrophe of major proportions might be to not open right away. A small business owner might reopen the business, and find that all their customers have moved away. Balance this against the idea that a firm that does reopen as quickly as possible might perform better in a hazard's aftermath, given less competition and increased need.[61] The choice is difficult because there is no clear answer; each business owner has to evaluate the business's unique circumstances and attempt to make sense of the situation.

New Orleans was filled with businesses, government agencies, and individuals being asked to make tough choices; the support networks that might have helped them to make those choices were not available to the extent necessary to help prevent some of the problems that occurred. The bar for radical and quick adjustment for survival can be set high in the days and weeks following a major hazard event. For businesses, a new normal might include adjustments to business plans, changes in how businesses market themselves, different customer bases, or perhaps offering different products or services. There is the question of whether the business's employees pre-event are even still available to work; it cannot be assumed that they are—some may have left the area entirely or are trying to personally recover from the event. In New Orleans, society's threads had snapped, and community and compassion were rare commodities.

New Orleans serves on many levels as a study in vulnerability. Vulnerability that exists before a hazard event might be manageable from the perspective of the individual or business owner. All of the work, income, and expenses balance out and people manage to go on living their lives. But when a hazard enters the equation, people who were barely getting by can no longer manage. All the assumptions of daily life presumed that the inputs and outputs would continue as they had, but a hazard event adds the element of disarray to the equation, and uncovers failures on the part of communities and other levels of government to properly quantify and plan for hazards. The cost of society-wide disarray can be considerable for an individual or business, but even an individual-level disaster can throw off the balance in income and outputs. An analog would be major health expenses for a person who is just able to meet all his or her regular expenses. The new normal, absent some other changes, is that some expenses will simply not be paid, or debt will increase. For a business, either decision puts ventures at risk. For local government, the role to be played in such a scenario is one characterized by an even greater need for leadership and action than would be expected under less vulnerable circumstances. If government does not perform its role, the role will not be performed.

New Orleans and its surrounding areas are unique. The population is quite diverse and rich in heritage, but that richness is threatened by an increasing gap in fortunes. New Orleans's businesses, particularly its small businesses, find themselves in a precarious state. The diversity is not necessarily well-supported by the official structures of the community as an end in itself, in a way that encourages the growth and development of small businesses. This is unfortunate, as the local small business community could be, through greater focus on retention and expansion of existing business, a source of even greater strength and resiliency for the whole community. For individuals, the distinctive New Orleans culture is at risk and deserves defense from both local business and government entities. With all the threats that New Orleans faces, the city needs as many strong advocates as it can find.

The findings of the regression analysis, notably that institutional culture is significant in impacting business resilience, are telling and worth additional consideration. The creation and enhancement of local supportive structures for business retention and expansion, and insistence within local government of fair and transparent application of regulations in all cases, will help to ensure a more business-friendly environment. A close partnership between the local business community and local government at a high level is essential. New Orleans must be open for business and for real economic development, or it will find itself exactly where it has been for decades—a tourist destination, subject to the whims of the market and lacking the stability that high-wage industries bring. Citizens must be involved in these discussions and their input truly welcomed. It will be a fundamental redefinition and rededication to the spirit of the city, or it will not work.

Even with all this in mind, there are community level issues that still require resolution. New Orleans community has been wounded by the tragedy that occurred. Within the community and even within the local government, there are trust issues with which leadership must dispense before government can make the significant progress it hopes to achieve for the people of the city and the region. Vulnerable groups in the city have been disenfranchised by official and unofficial actions for many years, well before Katrina and in the years following, and these issues must be sensitively approached with respect by the whole community. Planners have commented on the difficulty of such work, but difficulty alone is not a reason to avoid such efforts. Vulnerable populations have a role to play in the city's future, and ideally, all should rise together.

The officials interviewed for this project share a positive outlook on the city's future born of an obvious love for the uniqueness of the place. One commented, "I would tell them it's a great city, it's a fun city to live in, that hasn't changed. It's probably the most unique city in the United States and we're seeing a lot of new people move to this city, a lot of the young smart people that have decided that this is the best place to make a difference so I see this city growing, but I also see it growing in a much more sustainable and a much smarter way than maybe what we had pre-Katrina and that foundation that's being laid right now I think will bode very well business-wise for those companies that decide to make this their home." It remains to be seen if this attitude represents a clear break with the past and a way forward.

In the next chapter, we examine Palm Beach County and its school district. Themes of accountability and institutional culture echo in Palm Beach County as they do in Southeast Louisiana, with different results.

# 6 Palm Beach County and Hurricane Wilma

Located in southeast Florida and the northernmost county in a three-county metropolitan area that includes Miami-Dade and Broward Counties, Palm Beach County represents a divergent approach to both business resilience and disaster management. Practically synonymous with sunshine, beaches, and golfing, Palm Beach County has a multibillion dollar tourist industry, but has also shown great interest in targeting high-growth industries for economic development. The county shares a flair for diversity and international culture with its county neighbors to the south, and is a colorful, bustling metropolis. In 2012, the population was estimated as 1,356,545, which represents 2.8 percent increase over 2010. The median household income was $52,951.[1]

As with the New Orleans case, this chapter opens with background for the case, which centers on the impacts of Hurricane Wilma. Results of statistical analysis of factors affecting business resilience results are presented for both Palm Beach County government and the school district of Palm Beach County, as both these institutions have a unique set of businesses in their vendor lists. Interview results are presented, and a discussion and analysis section concludes the case.

## HURRICANE WILMA

"Although Wilma didn't cause nearly as much death and destruction as Katrina, it was certainly the year's most disturbing hurricane in a purely meteorological sense. The records that it broke—minimum sea level pressure, rate of intensification . . . captured the storm's ability to dissipate power and to potentially cause catastrophic damage."[2] A Category 5 monster at its peak, Hurricane Wilma had the lowest central pressure on record for an Atlantic tropical system, at 882 millibars. The storm had weakened to Category 3 before it made landfall in the United States on October 24, 2005, and crossing the Florida peninsula mostly as a Category 2 (extremely dangerous) storm, it exposed the state from Monroe County and the Florida Keys to central Florida to a wide swath of destructive winds. The storm produced significant storm surge, up to seven feet above normal in the Keys, and ten tornadoes throughout South Florida.[3] The fact that the storm had lost strength from the immense characteristics it exhibited in the Gulf of Mexico should not be taken to mean that the hurricane did not have harmful impacts to South Florida commensurate with a devastating event. Wilma's diffuse core scoured the southern part of the state and caused extensive damage over a wide area. Wilma was also a killer: according to the National Hurricane Center (NHC), the storm killed twenty-three, with five in Florida. However, there is disagreement between news reporting and the NHC on how many were killed in total from direct and indirect causes; other sources claim anywhere from ten to thirty in Florida alone.[4] Hurricane Katrina occurring in the same year perhaps led to the lack of research on Wilma and indeterminacy in calculating the cost of the storm.

As with Katrina, Wilma's Saffir-Simpson rating only tells part of the story. A Category 2 or 3 storm that destroys metal, wood, and concrete power poles throughout an entire region, flips cars and tears roofs apart, disrupting business and personal activities for a period of weeks, is catastrophic in effect, wind speed or pressure aside.[5] Figure 6.1 shows an example of Wilma's effect on area transmission lines. The storm's impact currently places fifth (moving down a notch to account for superstorm Sandy) among all hurricane events since 1900 in terms of costliness; when adjusted for inflation, the storm's impact was about $20.6 billion in damage in 2010 dollars.[6] This is evidence

*Figure 6.1* Power lines down across Stormwater Treatment Area, Central Palm Beach County, photo courtesy South Florida Water Management District.

that the risk to the South Florida region from these storms is great and perhaps even increasing, and that vigilance is not only important—it is essential.[7]

Beyond individual-level damage, the true extent of the event is evident in the loss to the South Florida electrical infrastructure. Electrical service was, for the most part, eliminated by the storm for at least days to weeks or months, over the entire area, creating a daunting task for the utility company Florida Power & Light (FPL) that required additional support forces from other states and a herculean effort to return power to residents (see Figure 6.2). "FPL has never had so many of its customers out, not even when a relatively compact Category 5 Hurricane Andrew roared through Miami-Dade County in 1992." Wilma knocked down 10,000 electrical poles and disrupted power to 3.2 million people.[8] "Media reports indicate up to 98 percent of South Florida lost electrical service."[9] In the FPL chronology of events, it is mentioned that Wilma impacted "about 60 percent of FPL's 35 county service territory, or about 22,000 square miles."[10]

While 77 percent of FPL customers had their powered restored by October 31, an impressive feat given the extent of the damage, customers in Pahokee, Belle Glade, and South Bay had to wait until well into November to see their power restored.[11] These western Palm Beach County communities are known for severe unemployment and large populations receiving food stamps.[12] They are also far-removed from the heavily populated areas nearer the coast; this further complicates power restoration activity and evades a more consistent reach for government services.

*Figure 6.2*   Electrical workers restore power after Wilma, October 26, 2005, FEMA/Jocelyn Augustino.

While Katrina was a flooding event in New Orleans, Wilma left mostly wind damage in its path through Florida. The winds did tremendous harm to the area's roofs, leaving behind a patchwork of blue tarps covering storm scars for months and even years, in some isolated cases. Homes and public buildings were damaged.[13] Streets were impassable following the storm and traffic lights were disrupted.[14] Public housing was rendered uninhabitable in Delray Beach.[15] Some people were still not back in their homes, as of 2012, due to problems with contractors.[16]

Palm Beach County had endured a three-hurricane season in 2004, and Katrina had previously dealt a glancing blow to South Florida earlier in the season; Palm Beach County had engaged in planning and fortification against the threat when Wilma loomed. Before the storm arrived, Palm Beach County had mobilized with both state and federal resources to assure that it was ready when the storm finally did come. On the Wednesday before the storm arrived, the county and its municipalities participated in a conference call, with police, fire, and emergency operations officials; officials placed orders for trucks of ice and ready-to-eat meals.[17] Arrangements were made with state and federal resources in the event that local resources were exhausted. Given lessons learned from the storms of 2004 in Palm Beach County, it was clear that certain needs had to be addressed in advance. According to one report, Federal Emergency Management Agency (FEMA) had sent "100 truckloads of ice, 100 truckloads of water and 30 truckloads of meals to federal staging areas in Homestead and Lakeland, along with Urban Search and Rescue Teams and Disaster Medical Assistance Teams" by the Saturday before the storm.[18]

Residents, familiar with hurricanes and having already dealt with Hurricane Katrina's brush with the area earlier that year, began prepping with the usual "plywood, ice, gas, water" several days prior to Wilma's arrival.[19] Residents understood that there might be no power and some, choosing to rely upon themselves, bought generators.[20] The Miami Dolphins rescheduled a football game to the Friday before the storm,[21] with the coach of the team suggesting that it "wasn't their first rodeo" when it came to dealing with hurricanes. Notices went out in media that schools, government, and other community organizations would be closed Monday and Tuesday, as one would expect.[22] The response from the community in preparation could be described as ordinary—routine, but competent.

Then the storm hit. Speaking personally, being in the area during the storm, Wilma's arrival was frightening; torrential downpours had pooled on the balcony of my condominium unit on the third floor of the building in Oakland Park, and wind forced the water under the sliding doors into the living room. The storm damaged our roof, all the while howling an unnerving roar; because of the leaks our roof, like many others in South Florida, needed repair. Cellular service was spotty and cell towers often did not have battery backups, so it was hard for my wife and me to contact relatives. The aftermath, with no power and no shortage of discomfort, was every bit

*Figure 6.3*   Hurricane Wilma damage at Port St. Lucie, 30 miles north of Palm Beach County; Wilma's damaging winds covered all of South Florida, photo courtesy South Florida Water Management District.

as unpleasant as one might expect. We were lucky that we still had water service. The news reports started to arrive of broken glass from skyscrapers in downtown Fort Lauderdale and Miami, to the south of Palm Beach County.[23] Trees were downed throughout the area (see Figure 6.3).

Disruption was the rule of the day. Governments and schools remained closed for many days. Wilma had interrupted the course I was teaching at Florida Atlantic University, and I had to become more creative in providing materials to students, since some class members lacked Internet access, not to mention the basic necessities for daily life. In the days after Wilma, citizens were frustrated by not only a lack of power, but limited water, gasoline, and ice. Long lines formed and irritation grew. Many retail establishments lacked generators and were unable to operate, leaving fewer options. The tendency at the time was to blame FEMA for the lack of response, but Governor Jeb Bush had another view, which recalled earlier warnings by local officials: "People had ample time to prepare. It isn't that hard to get 72 hours' worth of food and water."[24] For its part, local government had some confusion along the way, as well, in that they "prematurely announced distribution sites and times, causing crowds to gather hours before any supplies arrived . . . there simply was not enough ice, water and meals ready-to-eat . . . or it took far too long to get the supplies to the proper places."[25]

My wife worked in Weston, Florida, at the time and reported to work just days after the storm. While Weston started to get back to normal, other cities did not fare as well. While all places were not as quick to recover, businesses were still highly motivated across the board to recover from the disruption. The pieces were in place to spur recovery. Some banks were operational[26] just two days after the storm strike. Two days after the storm, the economic development organization in Miami-Dade was conducting a business needs assessment survey for local businesses that had seen damage from the storm.[27] The survey conducted by this group had the intended purpose of assisting "Federal, State and local officials [to] direct resources to the Miami-Dade County businesses most adversely impacted by Hurricane Wilma." Broward County government had begun calling certified small businesses in areas of work most appropriate for disaster recovery, to see if they were available to work; it later took out a full-page newspaper advertisement suggesting that residents use local firms for their recovery needs, to provide work to these firms, keep money local for maximum economic impact, and quell the use of unlicensed contractors. A bridge loan program was activated by Governor Jeb Bush on November 2, 2005.[28] A month after the storm, an article, written from the perspective of marketing the area to firms from outside the region, encouraged readers not to focus too much on the storm, "because . . . the hurricane didn't muster much notice outside the area."[29] In Miami-Dade, as one example, an attitude representative of the South Florida mindset appeared in the aftermath: "To site selectors, the message will be upfront: 'Yes, Miami had a hurricane, however we've rebounded well and we had business continuity.'"[30]

In hindsight, South Florida did mostly appear to "have its act together" when it came to responding to this particular storm, even with the occasional problems. Certainly, there were instances of opportunists that used the disaster for fraudulent personal gain in subsequent years, but for the most part, the local government response had the general feeling of being fairly competent, as were the preparations by interests at various levels of government, other community organizations, and the citizens themselves. We might notice just from the reports in the aftermath that the storm dealt an expensive blow to the area, but the business community rallied quickly, using networks and avenues of support to reopen. Institutions—local governments and public-private partnerships in the area—made being open for business a priority, and so the rebound time was very short. It was so short, in fact, that people from outside the region might not have remembered South Florida had this hurricane at all—it dropped "off the radar," as one article put it. South Florida has hurricanes and handles them, and this was "just another one."[31]

Local government institutions had a major role in this success, and the trappings of resilience that were exhibited after Wilma are worth considering. Palm Beach County is also a learning county when it comes to disaster events, in a high-traffic area for hurricanes, and has capability for learning from its mistakes and reducing vulnerability. It also appears that a sense of community also influenced the overall resilience experienced. Residents, and employees of local government, stayed in the area and were accessible for the response and recovery. Wilma was natural hazard event, but it was not a disaster in the literature's sense of that term. From the perspective of the business community, government, and individuals, Wilma was difficult at the time, but in hindsight all did a fairly capable job of managing the event and is aftermath.

## COUNTING THE PERSPECTIVE OF THE BUSINESS COMMUNITY

Referring back to the conceptual model for considering business resilience, business resilience is evidenced by the months it takes for a business to return to normal (recovery), and the capacity of the business in the aftermath of the disaster event. These two ways of thinking about resilience in the business community are thought to be influenced by four factors. We evaluate the potential impact of size of disaster, in terms of both damage to the business and the impact of personal damage on the business's recovery; the role of institutions, using both procurement programs and small business programs as variables that might explain business resilience; institutional culture, reflecting how the internal institutional environment may be impacting community resilience outside the institution; and vulnerability, as indicated by minority or gender status of the business's owner(s), whether or not the business had insurance, and the size of the business.

The surveys for Palm Beach County and the School District of Palm Beach County had response rates of 60.2 percent (571 responses/947 click-throughs) and 54.9 percent (237 responses/432 click-throughs), respectively. The approach to surveying mirrored that used for the New Orleans metropolitan statistical area (MSA) case, with the notable difference that both Palm Beach County and its school board were able to comply with my requests for information, including vendor lists. These lists were used to conduct the study.

For Palm Beach County, 74 percent of the firm owners responding to the survey were white; 9 percent were African-American, and 11 percent were Hispanic. 58 percent of firms were owned by males; 27 percent were owned by females, and the balance held by coownership or other arrangement. The split was similar for race for the school board, with gender of owners varying slightly: 62 percent male, 22 percent female, and the balance held by co-ownership or other arrangement.

When asked about contracting practices of the county and school board, business owners were more positive or noncommittal than negative, in contrast to perceptions of public contracting in the New Orleans MSA. Considering agreement with the statement, "When a firm is awarded a contract with local governments, it is based more on what they know, than who they know," 27 percent of firms (115) indicated that they disagreed with the statement, less than the 55 percent in New Orleans, though the greatest subset of respondents were ambivalent about the statement (136, or 32%). 134 firms (31%) agreed with the statement, "I am treated fairly on contracts with local governments, by local government staff," greater than the 18 percent that did not agree. In response to the statement, "The actions of lobbyists are a decisive factor in the award of contracts with local governments," 95 firms agreed or strongly agreed (23%); on balance, firms either felt that decisions were made fairly or did not express an opinion about the matter. The numbers for the school district were even better, with 40 percent (65 respondents) believing that contracts were awarded based on what respondents know; 64 percent (103) believing that they are treated fairly by district staff; and only 7 percent (11) regarding lobbying activity as a deciding factor in contracting awards.

As with the New Orleans MSA case, we review the responses of firms to questions about procurement following the event. Firms responded as follows to the following three separate statements, capturing elements of local procurement quality in that context.

1. I had access to government contracting opportunities that were related to hurricane disaster recovery.
2. For disaster contracts, the local governments protected the interests of local businesses in securing disaster recovery work.
3. During the hurricane recovery process, the award of recovery contracts helped to sustain businesses in my community.

Responses are shown in Table 6.1:

The large number of "neither agree nor disagree" and "N/A" responses in the Palm Beach County and School District of Palm Beach County cases are noticeable. This is much different than the New Orleans MSA case, where the negative responses (strongly disagree, in particular) are markedly more pronounced for these questions. Here, the responses expressing indifference are more pronounced.

Let us briefly examine the results of the regression analyses[32] conducted for each of the two business resilience dependent variables, for each of the two sub-cases within Palm Beach County, shown in Table 6.2:

We begin by examining the model for the dependent variable, months for the business to return to normal, for the Palm Beach County data set. We notice that the range of the business damage, even with the square root transformation, is much larger than in the case of New Orleans. The mean months to normal figure is much lower in Palm Beach County than it is in New Orleans (1.403 months in Palm Beach County, compared with 3.279 in New Orleans). Mean aftermath capacity is higher in Palm Beach County than it is in New Orleans (6.4 to 5.06 on a 10-point scale). Among businesses that participated in this survey, firms in Palm Beach County appeared to sustain less damage, which may be expected given that Wilma was a less catastrophic event than Katrina was in New Orleans.

In comparison with large businesses, small businesses are less likely to be insured for business interruption.[33] While not optimal, such a relationship might be expected within the model.

For the Palm Beach County data set, using time to return to normal as the dependent, two components of the model that are statistically significant

Table 6.1 Palm Beach County and School District of Palm Beach County, responses to questions centering on disaster procurement quality

| Question | Strongly Agree | % | Agree | % | Neither agree nor disagree | % | Disagree | % | Strongly Disagree | % | NA | % |
|---|---|---|---|---|---|---|---|---|---|---|---|---|
| **Palm Beach County** | | | | | | | | | | | | |
| 1 | 6 | 1% | 23 | 5% | 66 | 16% | 77 | 18% | 113 | 27% | 140 | 33% |
| 2 | 8 | 2% | 13 | 3% | 167 | 39% | 30 | 7% | 59 | 14% | 146 | 35% |
| 3 | 7 | 2% | 16 | 4% | 138 | 33% | 33 | 8% | 83 | 20% | 147 | 35% |
| **School District of Palm Beach County** | | | | | | | | | | | | |
| 1 | 2 | 1% | 8 | 5% | 47 | 29% | 28 | 18% | 22 | 14% | 53 | 33% |
| 2 | 0 | 0% | 3 | 2% | 77 | 48% | 7 | 4% | 8 | 5% | 66 | 41% |
| 3 | 1 | 1% | 9 | 6% | 59 | 37% | 11 | 7% | 18 | 11% | 61 | 38% |

*Table 6.2* Comparison of significant regression results among cases, Palm Beach County

| | Measures of Business Resilience | |
| --- | --- | --- |
| Study Area | Months to Return to Normal (recovery) | Aftermath Capacity |
| Palm Beach County | Personal Damage affected Business Recovery***; Disaster Procurement Factor* | Institutional Culture*; Size of Business*** |
| School District of Palm Beach County | Personal Damage affected Business Recovery**; Size of Business* | Damage to business*; Personal damage affected business recovery*; Institutional Culture*; Gender of Owner(s)*; Firm was Insured** |

*p<.1, **p<.01, ***p<.001

are personal damage affecting the recovery of the business, and the disaster procurement factor. That personal damage affected business recovery at a very high level of statistical significance is consistent with the vendor database's high concentration of small businesses. It is reasonable for us to conclude that small business owner finances are closely tied to those of their businesses and that, in disaster situations, recovery time may be negatively impacted by personal financial concerns.

Of interest also is the statistical significance of the disaster procurement factor. This has a positive effect on the dependent, meaning that quality in the procurement institution tends to increase resilience by reducing recovery time.

The other governmental institutional factor, for the small business program, was not statistically significant in its relationship with this measure of resiliency. Vulnerability and institutional culture also had no statistical significance in affecting business resilience in this model. This might be the case because the small business program may have little effect on outcomes as far as recovery time is concerned. The small business program in Palm Beach County is a program that establishes goals for subcontracting on county projects. If the firms that participated in this survey did not find that their fortunes improved by participating in the program, we may expect that they will not ascribe benefit to recovery to the presence of such a program. Further, the program may not provide for as much opportunity for new or different firms to enter the process as prime contractors or subcontractors by virtue of their participation. This would reduce the chance the program would have of making an impact on business resilience.

Vulnerability and institutional culture also do not return statistically significant results in impacting resilience. We might expect some measures of vulnerability to lead to changes in community resilience, but we can find little evidence of that here. It could be that the response to the hurricane reduced the recovery so significantly among businesses that there was little opportunity for issues of vulnerability to have an effect. Size and insurance also do not appear to be impacting resilience in any significant way. This bodes well for businesses in this county, because regardless of size, if something will impact resilience, it is a measure of vulnerability. This is not to say that vulnerability does not exist in Palm Beach County, but in impacting resilience, it does not in this analysis.

Institutional culture, similarly, is not impacting resilience. Responses for the question of "More what you know than who you know" tended toward a higher mean response, which indicates that firms feel, on average, that they are being evaluated on merits (knowledge and skill) when entering a procurement process with Palm Beach County.

Having evaluated the first dependent variable, let us move on to the evaluation of the aftermath capacity dependent as a measure of resilience for Palm Beach County. Since measuring recovery time represents only one facet of community resilience for businesses, another dependent variable was used to tap this measure in the case of Palm Beach County, similar to the analysis for the New Orleans MSA. This second dependent variable, measured as the percentage of capacity the business had in the aftermath of disaster, was run as the dependent variable in a second OLS regression model. The rest of the model was kept consistent, and included independent variables for size of disaster (damage to business, whether or not personal damage affected business recovery); institutional considerations (for procurement, the survey question response for whether award of disaster contracts sustained local businesses; and for small business, whether the small business program helped certified firms gain subcontracts); institutional culture, using the response for the "more what you know than who you know" question; and vulnerability, including race and gender of firm owner(s), size of business, and whether the firm had insurance.

The results of this regression analysis show that institutional culture and size of business are statistically significant in explaining the aftermath capacity of firms. The connection between institutional culture and aftermath capacity is central to the analysis of this study. A positive institutional culture is an important component of improving community resilience, because a focused local government effort can organize other partners and organizations within the community and provide for supportive networks that greatly expand a community's ability to respond in disaster situations. Institutional culture is a positive indicator here—it aligns itself with increasing aftermath capacity. The expression of the question—that it is more what than who a firm knows that leads to success—is a positive expression of ownership and fairness in the system, and is likely informed by a confidence in the local government institution.

Size of business is also significant, but at a higher level of statistical significance than institutional culture. Firms of larger size were more likely to show greater aftermath capacity. This is in sharp contrast to the findings in

the New Orleans case, but more consistent with some aspects of the literature that suggest small businesses have greater trouble responding to a disaster, in the initial response phase (what this indicator measures) and beyond.

Next, we examine the School District of Palm Beach County, using results
obtained from surveying vendors in their registered vendor database. The
school district has a separate vendor database, and makes purchases separate and distinct from Palm Beach County. The type of businesses in the
education-oriented database of vendors are different than those found in the
vendor directory for the county, where firms are generally oriented toward
providing services to a government with a broad mandate, including more
construction firms and licensed professional trades.

The school district vendor database showed less damage on average than
the results of the survey for Palm Beach County government. Aftermath
capacity was also higher. The proportion of minority vendors among respondents was lower in the school district sample (0.13 compared to 0.22 for the
county). Gender variable proportions were roughly comparable (0.27 for
the county to 0.22 for the school district). Firms responding for the school
district survey were slightly larger, on average.

As with the review of the Palm Beach County data set, we see two statistically significant indicators for the first dependent variable, months for a firm
to return to normal, for the school district data. Personal damage affecting
business recovery and size of business in gross receipts are again significant,
though at a lower level of statistical significance. This is consistent with both
the previous data set and the disaster literature. The lack of statistical significance for other factors involving the school district as an institution may be due
to the small amount of purchases or the low net effect overall of the district's
actions as it concerns the larger market. It may also be due to a low number of
responses. There are differences between the Palm Beach County data set and
that of the school district, as far as the registered vendor communities are concerned; it is worth noting that the school vendor database has a much higher
concentration of educational vendors, whereas the county vendor database
has a greater concentration of construction and construction-related services
firms, which are larger and more dependent on disaster-related procurements,
and other aspects of public interface studied here.

We now examine the second dependent variable, percentage of capacity
in the aftermath of disaster, for the school district database. The results for
this regression model show statistical significance in a wider variety of areas
in explaining resilience as aftermath capacity than the first dependent variable, which explored months for a firm to return to normal after disaster.
We note that size of disaster, as both damage to the business and personal
damage affecting business recovery, institutional culture, and vulnerability
indicators—the gender of the firm owner(s) and whether the business was
insured—are all significant in helping us to understand variation in aftermath capacity. There is a difference in size of disaster, though, as the size of
the business damage has the opposite directional impact from what would

be expected—as size of disaster gets larger, resilience increased, and the relationship is significant. This could indicate aptitude among businesses for quick recovery and is probably dependent on the type of damage the individual business actually experienced. Personal damage affecting recovery showed the expected negative result. The institutional factors, for disaster procurement and small business are, again, not statistically significant here.

Both business damage and personal damage affecting business recovery are significant, for the size of disaster component. This is important because the school district vendor database represents a unique set of firms, specifically those engaged in education-related trades, and how they respond to disaster is a potential area of inquiry for further research. These businesses may not be heavily capitalized or have projects proceeding that make recovery easier, as would firms in the county vendor database that are in non-education-related trades. This may explain at least some of the difference in size of disaster affecting firms in this sample.

Aftermath capacity is also impacted by institutional culture as a factor. The finding that institutional culture is statistically significant in explaining aftermath capacity, while neither of the institutional variables is statistically significant for this sample, suggests that a different approach to explaining business resilience on the part of the institution may be in order. What institutions, if not purchasing or small business, might impact business resilience, in the case of the school district? Is the amount of purchasing and small business participation in those programs so small that it would lead to their impact on business resilience being insignificant?

Gender of firm owner is significant, as is whether or not the firm had insurance. A women-owned business in the school district, therefore, may be expected to have relatively higher aftermath capacity than a non-women-owned business. This is unexpected and not exactly in line with what the literature suggests about the vulnerability of women (and by deduction, enterprises run by women). However, there is suggestion that Palm Beach County has had a relatively high proportion of women-owned businesses relative to the other firms in its market area (Broward County and Miami-Dade County), and this prevalence of women-owned firms may be leading to a stronger base for enterprises with such ownership in the area. Business insurance is also important for this measure of business resilience. This is a common refrain in official exhortations for communities to be prepared for disaster, and the suggestion is that having insurance will allow individuals to rebound from the disaster impact more quickly. In the case of the vendors in the school district database, this appears to be true—the connection between a firm having insurance and their resilience is significant.

## Interview Results and Analysis

As in chapter 5, the qualitative analysis in this chapter relies on coding of interviews with officials in Palm Beach County and with the School District of Palm Beach County. The general themes fall into four basic categories, with the detailed codes being the same as in chapter 4:

1. Reducing Vulnerability
2. Increasing Vulnerability
3. Inclusion and Transparency
4. Lack of Transparency/Accountability

Tables 5.5 and 5.6 in chapter 5 highlight the major coding areas of each of the major discussion themes, and those are retained for the comparison analysis used in this chapter, for the Palm Beach County and School District of Palm Beach County. The qualitative analysis considered all field texts from both cases at the same time, in defining the codes from their initial conceptions to the creation of major themes that wove the interviews, and cases, together. As with Chapter 5, Table 6.3 that shows the incidence of the various codes and how they contribute to an overall sense of meaning for this case, through the creation of major theme areas with positive or negative connotations.

One interview was conducted with school district officials; the interview was a combined discussion with several staff members representing various departments.

As can be seen from Table 6.3, discussions with Palm Beach County officials are indicative of feelings of increased inclusion and transparency. Of the vulnerability categories, codes that centered on reducing vulnerability were dominant. Where lack of transparency and accountability was discussed, it dealt primarily with the small business program and its efficacy. One official, faced with the reality of a tight budget, questioned the program in the larger scheme of operating a county government:

> You get into a question in tough budget times is that a must have or a nice to have, which can apply to many of the things that the county does, but singling that out you get into more of a philosophical discussion about whether that is core mission of the county government in tough budget times. But getting past that fact is "Yes," they do, what I consider to be a very good job.

## SYNTHESIS AND DISCUSSION

Comparing the results from the statistical analysis to the results of the interviews, we see confirmation of the principles identified in the project's conceptual model. Size of disaster, institutions, institutional culture, and vulnerability are all seen to impact business resilience in Palm Beach County, in one or both study sub-areas, for measures of business resilience. Public officials in Palm Beach County made a point in discussions to identify why

Table 6.3 Incidence of major themes from qualitative analysis, Palm Beach County and School District of Palm Beach County

| | Reducing Vulnerability | | Increased Inclusion and Transparency | | Increasing Vulnerability | | Lack of Transparency/ Accountability | | Total |
|---|---|---|---|---|---|---|---|---|---|
| Palm Beach County (5 Interviews) | 17 | 10.30% | 107 | 64.80% | 6 | 3.60% | 35 | 21.20% | 165 |
| Palm Beach County Schools (1 Combined Interview) | 3 | 7.00% | 30 | 69.80% | 0 | 0.00% | 10 | 23.30% | 43 |
| Palm Beach County (Total) | 20 | 9.60% | 137 | 65.90% | 6 | 2.90% | 45 | 21.60% | 208 |
| *Average of Themes per Interview* | 3.33 | 9.60% | 22.83 | 65.90% | 1 | 2.90% | 7.5 | 21.60% | 34.67 |

processes work as they do, and how processes are designed to be fair and open. More than one interviewee mentioned that Palm Beach County has been a proponent of emergency preparedness, and touted the effectiveness of the post-disaster redevelopment plan. From an economic development perspective, the conversations focused on regionalism and the strength of the county and the tri-county area (with Miami-Dade and Broward Counties) for business. Comments made were consistent—Palm Beach County is oriented toward creating a business-friendly environment, attracting new industry, and growing existing business.

The process of requesting and receiving data with Palm Beach County was, from the time I requested a copy of the vendor database to the interviews, open and transparent. The interviews included frank discussions, especially when one considers the cross-case evaluation. What was impressive was all the work that had been done to be proactive about disaster situations. Disaster procurements, for example, had in many instances been handled well in advance. Contracts were in place for required materials, including fuel. An official commented, "We bought our own tanker [of fuel]. I'm doing the same thing [now, preparing new disaster preparedness contracts]. Now there might be some more interest because now I hear Broward County has followed our suit, I hear that Miami, so maybe there's more interest." Arranging contracts for fuel in a disaster, in advance, proved complex, but the official commented, "You know, we try to be proactive."

For needs not anticipated in advance, there was a system devised that was mandatory for purchases: staff went to local businesses first for products that might be available locally after a disaster. The staff seemed especially aware of the importance of local businesses and intent on keeping business local after disaster and the process reflected this thinking. Officials responsible for such activities were to play a primary role as part of the county's emergency operations team in the event of disaster, raising issues of local business to a high level.

One official commented on the subject of disaster procurement: "when Frances came, what we've found is that we worked together like a well-oiled machine. I mean it was absolutely wonderful. I moved into the EOC [emergency operations center]. I had a core group of purchasing staff that I took with me, maybe five buyers, and a manager included—they went with me, I set them up in a room. They did the [response] purchases—We brought our vendor lists . . . we brought phone books." In trying to get products that were needed, local vendors were shopped first: "it was our training and what I would do, it would come up on the screen, and my staff in the back room would see it as well and they would start calling existing vendors that we have and if we couldn't get a hold of them, we'd start with the phone book."

The county small business program is successful, in that it achieved the expected percentages for certified small businesses on projects with goals, according to officials. But advocating for small business on county contracts is, itself, complicated. One official noted, "[we are] very conscious of the fact that we want to advocate for [certified firms] but we also tell the small businesses, you have to be ready and prepared to do the work, make sure you read your contract, don't sign anything without reading it because once you put the name on the line you're held responsible for it. You know, there's only so much we can do for you as far as assisting you." Small business officials were aware that they were opening a door for opportunity, but were limited in what they could accomplish within the program. They tried to inform would-be participants in county procurement of this fact. "We sit down and make [the small business owner] think 'Is this something I really want to do?' 'Have I sat down and really thought this one out?' And the big thing we really try to push on them too is have they really developed their business plan?" The office also stressed the importance of firms having the appropriate occupational and professional licenses.

Interviewees at the school district appeared to be doing as much as they could to implement their assigned programs, in creative ways. They were aware of issues with their programs and how their offerings could be enhanced to provide for increased transparency and accountability, and importantly, these officials were already working on those matters. They also displayed an impressive level of knowledge about their work and responsibilities. One official pointed out changes to procurement systems that had created a competitive bidding environment: "it is so competitive where the bids are a lot lower and so competitive because a lot of vendors are disqualified . . . so it's a pool now that is being widespread." Insistence on standards drove the competitiveness.

Contracts that would have been previously renewed were being solicited: "we pretty much just automatically renewed those bids when it came time to renew them. We don't do that now because there is so much competition out there and we want to give everyone the opportunity to have, to respond to our bids and open up some of those doorways for people." The official mentioned an active protest system for bids, which was utilized frequently because of the scarcity of contracting opportunities. Firms that did not win a protest "go away . . . not happy, but they understand our decision." Fairness was given as a central principle of operation: "Our goal is to get the best quality, the best price for the district, but also hand in hand that we have open competition that everyone is assured of fair chance to be awarded this and we truly believe in that." This is consistent the findings suggested by the survey, that businesses found the environment competitive and that the staff treated vendors with respect.

Bringing this back to the research question, business resiliency in particular does seem to be impacted in Palm Beach County by institutions. We might reconsider rejecting too quickly the hypothesis that small business institutions matter, because there are a variety of programs—through governments and through community based organizations—that might yield different results. Strictly speaking, this is a very limited look at a program that would not find its fullest usefulness to the community it intends to serve in a time of disaster.

We see indication that size of the disaster and vulnerability are variables that play a role in business resilience. As a point for future study, it bears additional examination but the significance is there. We do not see statistical significance for institutions in the school district, but this is not due to the lack of trying on the part of district staff, given the interviews. It is quite possible that the extent of interaction just does not support school district procurement having an impact on general business resilience in the aftermath of disaster.

Of concern in the Palm Beach County case is the fact that Wilma was, as a Category 2 hurricane, not a storm of *devastating* proportions, even disregarding its considerable disruption and damage (Figure 6.4 shows a photograph of accumulated Wilma debris). It was an expensive event to respond to and recover from, but imagine for a moment the potential for much larger events in Florida. The state sits at a particularly busy location when it comes to tropical activity. It could be said that Wilma, as damaging as it was, spared the area in not arriving at its previously epic stature, for example, or spending more time traveling across the peninsula. Imagine the need for business recovery centers after a catastrophic hurricane. According to the

*Figure 6.4*   Hurricane Wilma debris in Lake Okeechobee, photo courtesy South Florida Water Management District.

County Business Patterns database, in 2011 there were 41,970 firms in Palm Beach County alone. Any current system of business recovery centers would quickly be overwhelmed in the face of so much potential destruction, and a hurricane event would not stop at the county's borders. Officials may be underestimating need and overestimating inherent ability to respond.

After the 2005 season, hurricane expert Bryan Norcross expressed pause at the tendency of some politicians to congratulate themselves on how they handled Wilma, when the threat of much larger events is so unmistakable. He correctly notes that even leaving one percent of the population unmanaged in an emergency event leaves at-risk populations without the transportation, shelter, and services that they need. He wisely counsels about the importance of professional emergency management, and that that inadequate "learn-as-you-go politicians . . . make bad decisions because (a) they have not taken the time to be trained in emergency management . . . and . . . (b) their instant decisions on what to say and do are often driven by political . . . (translate: wishful) thinking."[34] Even in the Sunshine State, which leads the way in many respects on hurricane preparedness, there is plenty of opportunity to learn from past storms and become more adaptable, schooled in emergency management principles, and become more resourceful in obtaining the critical skills that are often drawn upon in crisis situations. Florida runs hurricane activation exercises yearly, but from these, the feedback about what works and what does not work needs to filter back to individual staff. As thinking in crisis situations and working as a team in such scenarios are skills that can be taught and learned, communities need not wait for large events to become aware of their level of preparedness. Teams need to function in both regular and crisis modes.

Further, there is opportunity to engage staff at the individual level, to encourage sensemaking and problem-solving ability. It was apparent from the interviews that many of the staff felt empowered by their agencies to work within regulations to provide services fairly, and to make use of resources available. They also seemed to trust and enjoy working with one another. These are positive signs. In addition, allowing purchasing and economic developers a place in emergency management planning and operations, shows foresight for eventual response and recovery activity.

We leave Palm Beach County and head northwest to Minot, North Dakota, to see how the city's experience in responding to the 2011 flooding event adds to the discussion of community resilience and the roles to be played by local government.

# 7    Minot, North Dakota, and the Mouse River Flood

The 2011 Souris (or Mouse) River[1] Valley Flood is the greatest flood event in the recorded history of Minot, North Dakota, a city that has had its share of floods. When the sirens signaling evacuation went off at 12:57 PM on June 22, 2011, the menacing threat of massive flooding from the Mouse River became a tragic reality.[2] Real-time information about rainfall in Canada affecting river flow was not available, and this adversely impacted notification about the danger. Four days to evacuate suddenly became two days.

When the waters finally crested on June 26, 2011, the Mouse River stood at 1,561.72 feet above sea level, well above the flood stage of 1,549 feet. The raging river flowed at an unbelievable 27,400 cubic feet per second, which "would fill an Olympic-size swimming pool in less than 4 seconds."[3]

The flood resulted in the inundation of 4,800 structures, of which 4,115 were homes; over 11,000 people were forced to evacuate. The flood covered a distance of thirty square miles of the Mouse River Valley, filling 11,456 acres with waters two to fifteen feet deep. Flooding was not limited to Minot, although the flood event centered on the city in many ways. The devastation seemed to find its primary mark in the city (see Figure 7.1). Damage sustained in Minot to infrastructure and property reached over $600 million.[4] "That summer, 2,700 Minot residents lost their houses to flood waters; only 10 percent of them had flood insurance."[5] A quarter of the city evacuated.[6] At peak occupancy, the flooding necessitated the use of 1,958 Federal Emergency Management Agency (FEMA) housing units.[7] Parks and other public areas, libraries, and about 200 businesses were also severely damaged.[8] The city estimated that it had $1.09 billion in unmet need as of October, 2012.[9]

Minot faced the double impact of a flood of immense proportions, and the competing stress of an economy that had taken off in surprising ways due to growth in the local oil industry. The effect of the oil industry, in spurring remarkable growth in a city that had previously stagnated, had already created a significant challenge for local officials. There was recognition among the city's leaders that Minot would not only have to rebuild once the floodwaters receded, but for all practical purposes, envision and create a new Minot. Minot would have had to build anyway—there was not enough available housing to fit the demand in the first place. When the

*Figure 7.1*    Flooding in Minot, June 24, 2011, FEMA/Bill Brown.

flood hit, faults in existing infrastructure were made more apparent. The vulnerability that had been less obvious in a city with flat growth suddenly became a grave concern. A lack of affordable housing became a chronic issue. Decisive action on the part of local government to protect the community was needed.

While this flooding in and around Minot, North Dakota, represented a major event for the people and businesses of the region, the literature on the impact of the flood has been scant. Coverage from an academic perspective is all the more important in light of the potential lessons that may be learned for other municipalities in light of Minot's experience in responding to the hazard. While other hazard events of 2011 may now be appropriately described as disasters, given society's inability to properly address the problems, the Minot flooding was and is being quietly and professionally addressed.

## MINOT

Minot is a city located in the Drift Prairie region of western North Dakota. In 2012, the population was estimated as 43,746, which represents a 7 percent increase over the previous year, outpacing the state's increase of 4 percent over the same period.[10] The median household income was $46,687, placing the city between the cases covered here: Palm Beach County, Florida, at $52,951, and Orleans Parish at $37,325. While the median income is relatively high, the Mennonite Disaster Service notes on its Minot service page that "approximately two thirds of households live below national median standards with nearly 13 percent of the population living below the poverty line."[11] Still, the level of population living below the poverty line in Minot is 12.4 percent, below Palm Beach County (13.3%), and less than half that of New Orleans (25.7%).

The city hugs the banks of the winding Mouse River, which begins in Canada north of Weyburn, Saskatchewan, before it dips into North Dakota and winds its way back into Manitoba; it then joins the Assiniboine River, a tributary of the Red River. The international nature of the Mouse has created unusual concerns that further differentiate preparation and planning, response, and recovery efforts in Minot from the problems of the other case studies presented here. The case is distinguished by the unique position of Minot within a state that has shown itself business-friendly and adept at reducing unemployment. Minot is heavily associated with the current oil boom, and while there have been numerous benefits from the growth seen in the city, this growth nevertheless has impacted response efforts for the flooding.

Of North Dakota, it has been said that "there is virtually no unemployment . . . while the national average appears to be stalled at an agonizing 8 percent. Wages are up. Land values are up. The long nightmare of rural decline and outmigration has not only been halted, but reversed in a breathtaking way."[12] North Dakota's positive relationship with the private sector, including oil, is notable. There is a certain amount of balancing that goes on in the state with regard to the interests of business and the protection of the environment. The interests of the private sector are not taken lightly, due in no small part to the fact that the state owes much of its success to the private sector in driving its present economic growth and record of consistency when it comes to employment. At the capital Bismarck and in the communities, there is a pronounced focus on state's rights. People do not appreciate the input of outsiders when it has not been requested; they know well enough the needs of their communities to be able to figure it out for themselves.

North Dakota's successful energy program, Empower North Dakota, has helped the state to prosper even in a time of general economic downturn. "[North Dakota] is the second largest oil-producing state in the nation, trailing only Texas. In 2012, North Dakota produced more than 245 million barrels of oil and provided nearly 11 percent of all U.S. output."[13] At the same time, the industry is sometimes at odds with regulatory sentiment from the federal level. The local and state governments generally support industry as a source of jobs and revenue, and it is undeniable that a strong economy and superb job availability has resulted in popularity for industry-supportive views.

The oil industry has been seen as a great advantage for economic development in Minot, as well as other cities around the oil fields. Work in the Bakken formation, colloquially called the "Patch," which underlies North Dakota, Saskatchewan, and Montana (see Figure 7.2), shifted from Montana to North Dakota in the early 2000s, leading to lowered unemployment and substantial increases to state coffers.

The Bakken formation represents a reserve of petroleum with great potential in the view of many to solve a variety of national energy problems, while also responding to local needs, such as employment and funding for vital improvements to community infrastructure. The impact has been remarkable: employment grew by 35.9 percent from 2007 to 2011, and total wages surged from $2.6 billion to $5.4 billion.[14] Rents have increased dramatically in what were formerly sleepy small towns; hotel rooms are

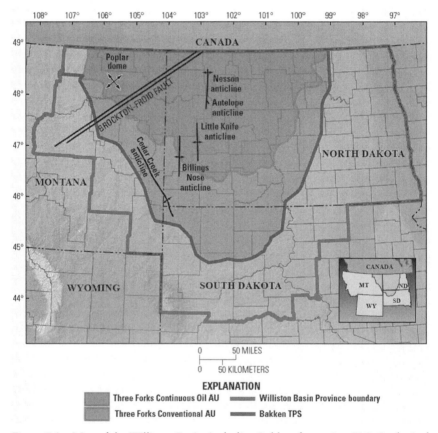

*Figure 7.2*  Map of the Williston Basin, including Bakken formation, U.S. Geological Survey, 2011.

tough to get in the city. Even with the potential for a boom-and-bust situation, environmental degradation, and societal impacts, it is difficult to say no to prosperity of this immediacy, especially when so many people do need the work.

Minot relies heavily on the influence of the oil industry in assuring its economic well-being as a community. The city has been fortunate in that the oil fields are an important resource in close geographic proximity and promise to continue to be of consequence for the foreseeable future.[15] Minot has seen the potential that industry has to assist the city in creating a bright future for its citizens and embraced the challenge of rapid development. Indeed, the impact of the oil industry likely had much to do with Minot's strong flood response and recovery. As a local official noted, "I don't want to minimize the impact of this flood, because it is huge. But this economic train that Minot is on, it's going to take a lot more than a flood to derail us."[16]

Minot has seen its share of boom and bust. Presently the city is in a growth phase, with the oil industry a primary driver of that growth. The prosperity that has accompanied the oil boom has been a great asset to the city and the region. Nevertheless, there have been some social concerns as a result of the change. Minot has gone from a relatively stable population to a growing population. Housing has become a top commodity, and affordable housing has been in short supply. This is a problem from the perspective of vulnerable populations, including the elderly, who are often long-term residents of Minot. The shortage in housing became worse as a result of the flood. The newer residents have different ways than the long-term residents which grate on people who have made the city their home for many years. In the local media, reports of rude cell-phone driving and texting have accompanied increases in prices for products and services due to this boom; this has been a point of concern for some long-time residents.[17] Suicides have increased in the state by 70 percent from 1999 to 2010.[18] Crime has increased marginally; while far from the crime epidemic standards of other cities, local government officials have had to remind residents of the need to lock their homes and cars.[19]

The oil industry in North Dakota has its share of detractors as it does elsewhere, from people openly opposed to development to those who just believe the growth has strained the ability of governments to respond effectively to new societal demands. The boom has placed stress on North Dakota's small towns, where workers live.[20] Even the simple increase in population would place an added demand on utilities and city infrastructure. However, the popular sentiment about the benefits to the economy and the need for jobs tend to make oil a good partner for North Dakota right now.

Local government has had to keep pace with the growth of the city. The city's total budget has increased from $86,530,294 for 2012 to $177,953,265 for 2013, an increase of $91,422,972 (105%). The budget includes major capital improvement projects, including sewer enhancements and downtown revitalization. The 2013 budget also includes thirty-one and a half new full-time positions, including a permanent public information officer.[21] The addition of a permanent public information officer addresses a point of critical concern for local government officials during the flooding event—they found that communications with the public could have been improved during the response to the flooding, and that the public information officer is a key position in any response effort. The importance of a public information officer (PIO) took some convincing; one official commented, "We thought we needed an equipment operator to run a snow plow far more than we needed a PIO, and honestly the last few years we did need that snow plow driver."

## LOCAL GOVERNMENT RESPONSE

The flood in Minot, like floods in Cedar Rapids, Iowa, and Grand Forks, North Dakota, has been seen in some quarters as an opportunity for the city to come back stronger, to build better and create an even more vibrant community.[22] Programs like Imagine Minot, a considerable undertaking to revitalize the city's downtown, have seen broad support from a coalition of funding sources. Imagine Minot looked to support a new vision of downtown by providing for not only expanded retail and office space, but also desperately needed affordable housing options.[23]

Minot officials have been able to persevere and overcome intergovernmental obstacles. For example, FEMA, through its hazard mitigation grant program, has awarded Minot $2,081,629, about 75 percent of project cost, for the engineering phase of a protection project for its water treatment plant,[24] with additional grant monies expected for the construction phase. About the project, Congressman Kevin Cramer suggested that it was "an example of rebuilding in a way that makes the city of Minot more robust against possible future flooding events."[25] Minot has also been successful in lobbying for support through grants from other agencies. Minot has gone after funding from a variety of sources, and gotten it.

A consistent theme in the city's response to the flooding event has been communication and keeping the door open to public participation and involvement in recovery efforts and planning. Local government has acted in a manner that invites the perspective of the citizens in responding to the flood and potential for future flooding—city hall has allowed for an open environment where feedback is solicited and addressed. Even negative comments are heard in a sensible way that encourages continued involvement from citizens. Public participation in governmental and regulatory processes can sometimes be complicated and delay governmental action. In spite of any complexity that such processes might cause, actively involving the public in a hazard situation reduces the potential for pushback against recovery and prioritization efforts. This is another reason that devoting adequate resources and staff to disseminate public information is so crucial. Further, actively engaging the public creates a mindset in the community of involvement and ownership—such advocacy turns residents from customers to citizens in the most optimistic sense of that term.

One of the reasons that recovery in the park system was slow was that the flood protection plan had yet to be settled. Determining what areas would be left open to protect against future impacts from a flooding Mouse River put a damper on some efforts to move forward to make places as they were.[26] However, when there has been a clear way forward with projects, efforts have moved quickly and involved the community in unique ways.

For example, Minot's park system has been a point of pride for the community and in the wake of the flooding, working toward recovery in the parks allowed residents a way to become involved through volunteerism and fundraising. Minot's Oak Park, which was damaged by the flood, remained covered with debris and unusable for some months after the flood.[27] Nevertheless, in October 2011, the park was named America's Favorite Park in a Coca-Cola contest, which netted the city a recreational grant for $100,000. This was a community-driven effort that saw an opportunity to gain some much-needed funding for the park in a time of crisis and was successful in obtaining it.[28] Minot's Roosevelt Park Zoo, devastated by the flooding in 2011, reopened on May 4, 2013. As it worked to reopen, the zoo was helped by a Minot State University strategies class, which held a costume dance with a jungle theme as a fundraiser.[29] Staff and guests from the nearby Staybridge Suites hotel volunteered to help with painting benches and painting trash cans.[30]

The city has also involved residents in long-term planning activities through open houses. According to the *Minot Daily News*, "Residents who took part in [an] open house . . . to discuss the city's comprehensive plan expressed concerns about what the city will look like in 10 years, were encouraged by plans to revitalize parts of downtown, and wondered how city planners would rein in the exploding residential and commercial growth in the area."[31] That citizens are interested in the city's future and that their feelings are being heard is encouraging.

The city has looked to actively engage neighborhoods through a program of revitalization planning. "The goal is to involve both neighborhood newcomers and people with long ties to their areas. The project aims to connect neighbors and get them thinking about ways to improve their communities."[32] The city has also made efforts to involve residents in the Emergency Support Function (ESF) 14, which includes identifying key recovery priorities. In Minot's case, these included "suggestions for residents on affordable housing, a Souris Valley greenway for parks and recreation, zoo recovery, a transportation study, bicycle and pedestrian paths, flood protection, downtown revitalization, redevelopment along the Burdick Expressway corridor from Broadway to the fairgrounds, business development, expanding the retail base, education, health services and historic preservation."[33]

Not everything about the response and recovery has gone as planned or quickly as desired. About community recovery, Minot Mayor Curt Zimbelman commented, "I felt we would come along quicker than some communities, and I think we have. I think we have come a long ways. Certainly when you are dealing with the public and all the different situations, I realize in many cases we are not moving fast enough, especially on the [home] buyouts."[34]

Part of the problem has been that federal assistance for mitigation has involved some major restrictions on how the funding is used. The city wanted to change the level of the floodplain, to access assistance for mitigation against future flooding, but doing so would require it accept requirements, like mandating that inhabited floors of buildings be one foot above the new floodplain guidelines; this would fundamentally alter large sections of the city.[35]

Community partners, nonprofits that assist the public generally and business-related interests, and faith-based groups have worked very closely, if not seamlessly, with local government. Granted, the parties do not see eye-to-eye on all issues, but the interchange between parties is robust, honest, and reasonably effective in getting at the issues facing the city.

While the city and partners are working together, the growth caused by the boom had some unintended consequences. The housing crunch that accompanied the oil boom expansion also caused some difficulties in responding to the flood. Flood volunteers were anxious to do the work of helping to get the community back on track, but with hotels and temporary accommodations being few and far between under normal circumstances, there were even fewer places to stay in the wake of the flood.[36] To their credit, responders did not allow this to obstruct recovery; they creatively developed Hope Village, "a temporary community that will house incoming volunteers. It includes trailers that provide sleeping quarters, bathroom and shower facilities, and a large dining tent that serves three meals a day. It is the culmination of efforts from a coalition of local faith-based and community volunteer organizations."[37] Hope Village, opened in April 2012, is one of the more unique aspects of the recovery in Minot because it involves people of all faiths working toward a common goal. Differences in faith take a back seat to assisting people, even as they strengthen the community's response through diversity of belief. Hope Village streamlines processes for those in need and for volunteers to enable a seamless system for dispensing help of all kinds—getting people back into their homes and on with their lives.

The city has begun to come back in important ways. The theme of a summer of hope took center stage in 2012, echoing the call of Hope Village. Ubiquitous "I'm Coming Back" signs were used by residents to convey their hope, and perhaps even a little defiance against the flood. These are now being revised to say "I'm Back." Transportation infrastructure is returning as a priority. Amtrak put a $500,000 into refurbishing the city's train station.[38] A major airport expansion project is on the horizon to replace the city's current, overextended airport. The city is beginning to reflect what will eventually become its future. But the international river that flows through the city continues to be a threat.

## AN INTERNATIONAL RIVER

The international nature of the Mouse is where part of the concern lies with responding to a flooding event; the International Souris River Board takes a primary role. Minot is conspicuous in its physical vulnerability because it stands astride the river and is prone to flooding when there is great snowfall, or rain of unusually high volume. Real-time runoff information was a major factor in the city of Minot being less well-prepared in the flooding event of 2011. In 2011, citizens were forced to evacuate on two occasions, given the uncertainty of the flood prediction. That the river did not crest at a higher level in 2011 was attributed to quick action on the part of Canadian officials to redirect waters to reservoirs before they reached Minot.[39] It is clear that cross-border discussions about the river and reporting potential threats were, in 2011, insufficient to fully protect the city's population.

Even into 2013, reservoir management in Saskatchewan and communication of the state of potential flows into the Minot area remained a concern. A March 2013 conference call gives a strong indication that talks between Canadian provinces and North Dakota officials were still emerging; Saskatchewan Environment Minister Ken Cheveldayoff indicated that "the purpose of the call was to open the lines of communication and to let officials on both sides of the border share information and then agree to share information going forward." It is not clear from this statement that the cooperation and collaboration that could make an important impact in preventing future floods had been realized.[40]

There are differing reasons between Canada and the U.S. for paying for dams and control measures along the Mouse—Saskatchewan needs stable water supplies, and North Dakota needs flood control. Efforts made to coordinate controls between the two countries are not new; the current scheme operates under a 1989 agreement. Existing arrangements did not plan for a 500-year flood event, as the 2011 Minot flood has been deemed. Regardless, Minot has experienced flooding before and there might well have been some expectation that flooding of this magnitude was a possibility, given the momentous 1881 flood, and the threat of flooding and other times between the two large events. The excessive rain Saskatchewan was experiencing, and its translation into intensification of a potential flood event, was not adequately communicated. This imperfect alignment of events and scenarios, along with a lethargic level of communication about the pending threat, led to the devastating flood event of 2011.[41] If there is a failure in the case of

Minot, it is the failure to bridge the gap between officials in Saskatchewan and North Dakota to protect the public interest.

There has been some criticism of the capacity of the Water Security Agency of Saskatchewan, formerly the Saskatchewan Watershed Authority, to appropriately manage water resources and prevent the kind of flooding threat that occurred in 2011. Some of the interviews noted that the information was less than needed to make qualified decisions. At one point, Minot officials were trying to get information any way they could to protect the public: "our [city staff member] asked if . . . [there was a contact] in Saskatchewan to get some kind of an update on what had actually happened there, and he said, yeah, he thought he could do that. So there was another call that day . . . they knew that there was more than what had been forecast, more than what had been anticipated and more than they knew about at that point . . . but still the Weather Service didn't give them good data . . . looks like it could have 10,000 CFS [cubic feet per second]."[42] In exploring every alternative for information sources given the absence of official information, this early information, before the much worse predictions that would come in the following days, may have gotten some movement in preparations that resulted in lives being saved; the city, which was already closely monitoring the situation, was making the best use possible of available information. So what happens when even that is not enough?

## Methodology

My handling of the Minot, North Dakota, case deviates from the New Orleans metropolitan statistical area (MSA) and Palm Beach County cases in an important respect—no new quantitative survey was conducted as part of the case. The reasons for this are threefold. First, Minot does not engage in public procurement in a manner similar enough to the first two cases to warrant a business survey of the type used for those portions of the study. Second, Minot has a much smaller business community (2,093 businesses in the Minot, ND, Micropolitan Statistical Area as of 2011[43]) than either the metro New Orleans or Palm Beach areas, so comparisons based on a survey would be confusing at best. Finally, Minot, through its community partners, has commissioned surveys of post-flood business recovery that are of commendable quality. A review of the results is included here to account for triangulation. Discussion of business recovery, given this secondary data and coupled with findings from the interviews, is presented in the discussion section of this case.

As with both the New Orleans MSA and Palm Beach County cases, the analysis of the Minot, North Dakota, case used grounded theory to view the experience of the event through the eyes of those who were affected by it and now called upon to plan the city's future. The review here of the Minot case is informed by a particularly rich and rewarding interview process, which yielded subsequently strong insight through application of grounded theory techniques. Development of comprehension of the unique experience of Minot was achieved through an iterative interview process and constant comparative analysis of the results. Where the strength of the Palm Beach case was in the strong response from the business community to the survey, the strength of the Minot case was the openness of public officials in Minot in casting a light on the city's response to the most extraordinary hazard event in its history, and a willingness to speak frankly and consider how local government and its partners had learned from the experience in ways that could help others. As with the cases of New Orleans MSA and Palm Beach County, Florida, I used the open—axial—selective coding approach in grounded theory.

A total of ten interviews were conducted for the Minot case study. For the eight in-person interviews, I recorded the discussion on a portable MP3 recorder after obtaining permission to do so. I then ordered the recordings transcribed by a third-party transcribing service (FoxTranscribe, now Rev.com), mentioning the academic nature of the work and requesting confidentiality. On September 30, 2012, I emailed the interviewees a copy of the transcription and requested review and approval. Most responded favorably or had minor changes. One official responded to express concern and clarify some points. For purposes of disclosure of the methods used here, I included the transcription from this particular interview with this official in the development of grounded theory for the case only, and have not utilized any quotations from this interview. All quotations used here come from the remaining interviews. My interest in the case was principally to tell the story of Minot and its flood response, and gather lessons that may be of interest to others. We now turn in more detail to the interview process for the case.

## Interviews

The interviews conducted for this project were held with a variety of public officials representing local government, finance/procurement officials, officials representing business interests, and nonprofit agencies/faith-based relief organizations. I first contacted officials in Minot via email in December, 2011, to gauge interest in speaking about the flood response, as I had seen no coverage in the literature and very little in the media after an initial National Public Radio report. I then spoke to two officials from the Minot area by telephone in January and February 2012. I prepared an initial set of questions and had a lively discussion with these officials about Minot and the response that was indicative of a need for additional work and research in the area. I found scheduling interviews with officials in Minot to be quite different from either the New Orleans or Palm Beach cases; rather than questioning my motives or interest in study and research, the request was received warmly.

I subsequently visited Minot to conduct interviews in person from July 23 through 25, 2012. I arrived a few days before the interviews and toured the area, including a drive into Saskatchewan and along the Mouse, so that I could gain a greater familiarity with not only Minot but also the wider region and its geographic features.

For the interview questions in Minot the question lists were modified slightly for each interview group to best serve the potential content of the discussion, to make the questions relevant to the work of that particular group, while keeping the thematic elements of the interviews as constant as possible. I began with a list I used for the telephone conversations (Appendix B, Section 2, list A), which had some commonalities with the questions used in the previous two cases. I quickly realized that with grounded theory I needed to listen to the responses and allow the interview responses to guide my understanding, even to challenge my existing grasp and to create new impressions from the perspectives of those that had gone through the experience of the Minot flooding. In undertaking this new research, I recognized that Minot presented an opportunity to gather additional information through that could potentially assist in further defining the role of faith-based groups in disasters, as well as the impact of other more community-based factors on the response/recovery effort. These considerations are reflected in the questions I asked in Minot (Appendix B, Section 2, lists B through E), and why they are somewhat different from those used in and around New Orleans and in Palm Beach County. As with the other cases, I refer to officials generically, and their names are withheld for consistency with the remainder of the project.

## Analysis

### Review of Business Resilience through Surveys

According to a flood recovery survey conducted by Ondracek and Witwer in 2012, for the Minot Area Development Corporation, 853 Minot businesses were contacted for their perspectives on recovery and how businesses have been impacted. 42.2 percent of the solicited surveys were received. About 97 percent of businesses were back in operation. The remaining 3 percent included businesses that had closed permanently or moved away from Minot. 62.3 percent of Minot businesses report that they are not fully staffed. The authors indicated that about 1,544 more employees are needed by Minot businesses.[44]

Ondracek and Witwer had previously conducted a survey on the impacts of the flood on the economy, in 2011, and found that the results showed consistency in questions asked between the two surveys, such as how many employees had left the area (reported 843 in 2011 survey, 827 in the 2012 survey). Businesses in both surveys had reported few closures, though like the other cases, where there was significant impact to a business, the process to return to normal can take a relatively long time.[45] While businesses have largely returned to normal in Minot, they continue to be impacted by the need for employees. Given that there were a significant number of employees that left the area after the flood, the already considerable need for staffing was made more obvious.

Minot businesses were assisted in the road to recovery by an active area Chamber of Commerce. The state also sought to assist businesses in getting back to normal, through the creation of a loan program: the "State Legislature approved $50 million to fund the Rebuilder's Loan Program, directing the State Bank of North Dakota (SBND) to offer fixed, one-percent disaster assistance loans of up to $30,000."[46]

It is worth examining the MSA Business Patterns information for the Minot Micropolitan Area to note the nature of the business community and trends that have been seen. About 86 percent of the 2,093 businesses in the Minot Micropolitan Area employee fewer than 20 employees.[47] About 55 percent of the businesses in Minot (1,141) employ between one and four employees.[48]

Table 7.1 represents Minot's top areas of business specialty, by number of total establishments.[49]

The category mining, quarrying, and oil and gas extraction includes just fifty-three businesses, but these businesses employ 1,459 with an annual

*Table 7.1*   Minot Business Specialties, by NAICS

| Industry Code Description | Paid employees for pay period including March 12 | Annual payroll in 1000's | Total establishments |
|---|---|---|---|
| Retail trade | 5,239 | 136,016 | 317 |
| Construction | 1,530 | 94,844 | 280 |
| Other services (except public administration) | 1,216 | 26,320 | 206 |
| Accommodation & food services | 3,830 | 48,347 | 184 |
| Health care and social assistance | 4,673 | 212,849 | 160 |
| Finance & insurance | 1,688 | 64,712 | 157 |
| Professional, scientific & technical services | 665 | 36,313 | 136 |
| Transportation & warehousing | 766 | 45,217 | 135 |
| Wholesale trade | 1,677 | 97,991 | 118 |
| Total | 26,281 | 1,006,946 | 2,093 |

*Source*: U.S. Census MSA Business Patterns (2011)

payroll of $125,068,000 (about $85,000 per employee on average).[50] The impact on the community from these high-paying jobs on other demand in other sectors of the economy is unmistakable.

We now turn to a review of the flood through the eyes of those that experienced it, as local officials or as community partners in the response/recovery effort.

### Review of Community Resilience through Grounded Theory Analysis of Interviews

In keeping with the "little logic" of grounded theory,[51] the discussion of Minot's local response to the 2011 flood here centers on how there is a story to tell in this case that has yet to be told. When the flooding first occurred, Minot was on the national stage for a brief time. When the floodwaters crested, and there was no death or grave injury, the national media in large part ceased covering the event. Among the people in the city, there were few outward expressions of drama beyond the relatively minor disruptions of houses to be de-mucked and otherwise cleaned, citizens needing to find a place to live until matters could be addressed, and quiet displays of the will to continue to work and actively engage one's daily life. There were some brief expressions of anger as the community worked through its grief, which were remarkably well-covered for being the non-stories that they were. However, the media simply failed to cover Minot's ascent after its collective hardship. Minot was quickly supplanted in the national consciousness by Hurricane Irene; when Irene arrived in August 2011, it took any interest in Minot from the national media with it.

To summarize the central premise of the community's response to the flooding in Minot, based upon the grounded theory exercise: Minot's response concentrated on the essence of the community working together, in a way that utilized the skills available to their maximum benefit for all concerned. This leads to a model of community resilience (Figure 7.3) that utilizes the original model of business resilience and expands upon it, and shows more of the complexity that may be at work in defining community resilience. Institutional culture is no longer shown as influencing only business resilience, as it influences the resilience of the whole community.

Resilience, the original dependent variable, is thought of in the Minot case as representing total resilience, including individual and business resilience and the resilience capacity of the public sector. The city, beyond support through procurement or small business, shows a close relationship between business, nonprofits, the public sector, and individuals in the community, and a tendency for all concerned to work together closely to solve problems and accomplish goals. This may be a function of the size of the community relative to the other case studies, but it may also provide a clue as to how agencies in larger institutions have a tendency to become mired in information silos, and may lose critical sensemaking ability as they lack engagement with other functional areas within the larger institution.

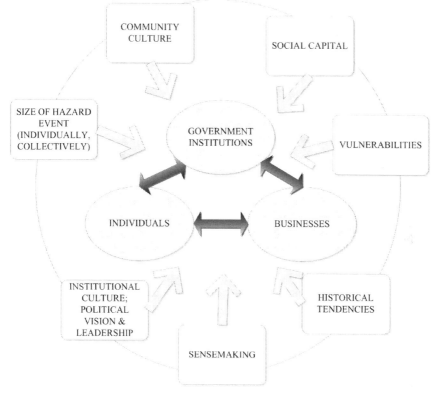

*Figure 7.3* Conceptual model of community resilience.

The major themes and representative codes for each are represented in two tables; we see codes related to major themes of communication/skill/ planning, and vulnerability. As seen in Table 7.2, the need for better communication in all respects was a frequent refrain in the interviews; vulnerable populations in the community were known to the officials being interviewed, and these officials seemed not only aware, but engaged in the provision of service to them.

In Table 7.3, we see the experience of the Minot flood as viewed by the groups contributing to community resilience—individual perspectives and the view from nonprofits and faith-based groups, identified collectively as "community," local government and assistance from other sources, including the federal government; and the business community.

The explanatory variables that are brought to bear upon resilience in Minot are, as in the original conceptual model, size of disaster, institutions, institutional culture, and vulnerability. However, the city has shown an active effort toward fostering inherent resilience, not just maximizing

*Table 7.2* Communication/Skill/Planning and Vulnerability themes in Minot

| Communication/Skill/Planning | Vulnerability |
|---|---|
| A public information officer is essential to response/recovery | Flood created special problems for the elderly |
| Planning to prevent future disruptions is essential; infrastructure considerations, flood planning | Event affected some neighborhoods more than others |
| There is a need not just to rebuild, but to plan for expansion. | Vulnerable populations closer to the river |
| Grant aid/skill set instrumental in gathering funding necessary for recovery | Social problems (e.g., driving intoxicated, domestic violence) |
| Information not out fast enough | Flood insurance needed, but citizens did not have/not required to have it; did not understand true risk/threat |
| Unclear/conflicting information | Mental illness and personal trauma from event |
| Should have made much better use of email, given that it is fixed and accessible via mobile. | Day care services needed desperately; lack of capacity |
| Had information that could have been used for follow up but did not use it | Volunteers need to take care of their own emotional well-being |
| National press coming in did not help; flood event was not adequately covered | Lack of affordable housing |
|  | Victim mentality and sense of entitlement among some only marginally impacted |
|  | In school system, flood impacted schools; temporary classrooms had to be used |
|  | Frustration and traumatic stress; flood response wore people down |

resilience in the adaptive sense when confronted with challenges. Whereas we explored institutions and culture through procurement and small business as a window on institutions in the first two cases, in Minot we examine the city and its partners collectively.

We begin first with the size of the disaster. The flood represented a huge loss to the community—a point shared by a number of the interviewees. The size and scale of the flood was overwhelming and created numerous problems for communication, allocation of resources, and other difficulties

*Table 7.3* Community, Government, and Business themes in Minot

| Community | Local Government | Federal/Outside Assistance | Business |
|---|---|---|---|
| Community working together | Having to choose between priorities/what to save | Problems with FEMA; Federal rules out of line with local needs | Business owners had to act quickly to save businesses |
| Faith-base as central to success | Staying in place and putting in long hours for weeks/ months | Help available, but not the help that is needed; restrictions on how assistance can be used | Whether housing was impacted also impacted business recovery, especially for small businesses, but also for employees of larger businesses |
| Summer of Hope theme | City supportive of community needs at all levels of government | Role played by Army Corps of Engineers was positive | Contractors can make more money elsewhere in rebuilding; this caused delays getting contractors |
| Volunteer help has been huge part of recovery effort | Supportive of nonprofits—recognition of skill and capacity in nonprofit/faith-based sectors | Agencies were able to work together at various levels of government | Business owners dealt with problems because they needed to be up and running |
| City not needing outside help; does not need government (esp. federal) help | Stress of flood aftermath—strain on public officials | Used common assistance application forms, which helped processing/efficiency | |
| Population with common culture/ relationships a strength | Recognition in the community that government acted strongly in the face of the threat; competent leadership | | |
| "Not a handout, a hand up." | | | |
| A positive feeling about the future | | | |

*Continued*

*Table 7.3* (Continued)

| Community | Local Government | Federal/Outside Assistance | Business |
|---|---|---|---|
| Good North Dakota work ethic | | | |
| Community leadership/decisions/vision from local officials | | | |
| Concern about flight of young people from North Dakota and affect on N.D. communities | | | |

that amounted to the creation of a "fog of war" scenario. One official commented, "Nothing this scale or this size is ever going to have everything go smoothly. There's a lot of people that compare things like this, a disaster, to being in war, because chaos is going to happen. It's going to ensue at some point. All you can do is try and control it as best you can." Regarding how decisions are made, one official commented, "I think the staffing, the knowledge, everybody makes the best decision they have with the information they have, right? Nobody gets up in the morning and says 'I'm going to screw something up today.'"

Some of the comments were grief-stricken and show the trauma experienced: "We have kids that when they hear sirens, because that's how they indicated that you've got to be out of the valley . . . the kids cry when they hear sirens." Others did not expose the raw nerve of the pain of the event; "I think everybody was just kind of dazed, walking in circles type of thing." For those that seek to provide comfort and assistance, it is arduous work; "It's a huge piece. It takes sometimes ten hours to go through to find one flood victim's needs. What do they need? Is it mental? Is it physical? Is it spiritual? All the different areas of a person's life are affected after the flood, they go through all of that." The city had an unmet needs committee that would meet and address these cases. They would hear these cases—there was so much loss that it speaks to the character and determination to assist others of this committee to keep to their task.

A number of the officials described the pain of the experience of dealing with the response and recovery in the days and weeks after the flood, and the trauma it caused them and others. The extent of individual-level problems was draining, and impacted other parts of the effort. One of the officials was still living in a FEMA trailer, which provided an extraordinary example of empathy. Transportation routes were destroyed or disrupted, which created challenges for those responding and simply trying to resume their lives (see Figure 7.4). Frustration, in cleaning up and moving on—just wanting to be done with it all and back to normal—was an important theme. "The morale of the community was just in the dumps. It was more or less washed away like so much was in our community . . . I don't think can't understand that until after it's done . . . I mean water in your attic. Everybody knew it was going to be a tragedy as they emptied their homes, but reality versus perception was long apart." The event clearly challenged people at their most personal levels, and wore them down—from citizens to the people challenged with the task of working to restore order.

The institutional experience in Minot is unique among the cases reviewed in this book. For one, the first thing I heard about the public officials in Minot was that they worked tirelessly in their response to the disaster, and that this was an intensely personal issue for everyone concerned. The lesson I heard more than once from the interviews was that local governments need to hit the ground running in responding to disasters. Planning for recovery occurs well before an event; being mindful of potential hazards

*Figure 7.4*   Highway 2, east of Minot, June 25, 2011, FEMA/Andrea Booher.

that threaten communities is not optional, and should not wait for another day. In Minot, there was no small amount of work being done to assist the community even a year in to the recovery process. This was a small, highly skilled, dedicated team working toward that effort. The capacity needed to respond to the event did not appear when the ESF for recovery was activated—capacity to respond was a trait that community leaders had already been developing.

The elected leadership was called upon to provide vision and articulate the way forward for the city. According to one official, "the mayor was as credible and as good as you could expect from anyone. He had been on the city council since 1978 and was elected mayor in 2002, so he knew our operation, he knew city government forward and backward, and he could communicate at a level that people could understand. He was very believable and would get emotional sometimes." This is no simple task and there is something to be said for the masterful ability of certain leaders to inspire confidence in a community when it is needed. In this case, competence and the ability to communicate clearly and honestly on the part of elected officials served the city well. Elected officials and administrators worked as a team; the difference a united front can make in raising the impression government makes in the public should not be underestimated.

It helped that the city was monitoring the threat from early in the year, even ordering an evacuation once the month before on threat of flooding. "Standard procedure would be that the disaster response comes from a combination of the county DES [department of emergency services] which is connected to the state DES and then the city, of course, as the event taking place through the county, but the principal affected entity was Minot so the community . . . the city government was deeply involved in obviously the defense of our community trying to protect against the flood event." When the threat level increased, meetings became more frequent: "We got to the point where our disaster group was meeting pretty regularly. If it wasn't daily it was almost daily, not because we knew that there was going to be a flood at that point but just trying to be prepared and making sure that we had things covered that needed to be done."

It soon became clear that the flood was going to be too much for the city to manage on its own. One official commented on the experience:

> *I mean we had fought for weeks. I can tell you it was the most humbling experience in the world for those of us who had been involved in the fight to know that we had lost the flood fight. We thought we were winning and when we lost the flood fight it was a humbling experience to suddenly be on the side of getting telephone calls from counterparts around the country saying, "What can we do to help?" And to be on the receiving end of that, it's not something that we're used to doing. We're much better at saying, "How can we help you?" than to have that come back on us.*

When the flood event occurred, the city's employees put in all the time necessary. "[City employees] were probably pulling 100 hours a week during the flood fight . . . before, during, and after. They were all full-time and then some . . . I mean, it's pretty obvious that they couldn't continue that because of the burnout rate would be way too high." City leaders had many responsibilities, some of them overlapping to ensure coverage.

Not everyone stayed to fight the battle before them in the wake of the flood. For one, the county emergency manager apparently left and never came back. "During the response there was certainly an amount of disorganization. Part of it is we're a part of Ward County and Ward County had an emergency manager. He went home on May 17 and never came back. There was not another emergency manager through the majority of the flood at the height of the battle. It wasn't as organized as it could have been." This is an unfortunate turn of events for any community, but particularly so for a local government that so needed the leadership of emergency management acting as part of the team. But at some level the loss of the emergency manager became representative of the broader challenges not only in this case, but in the problem confronting local government. If there is a message that can be conveyed, it is that each member of a response/recovery staff is needed to do their job to their utmost, even when others, for whatever reason, decide to look after their own interests. Even when all staff are not available, the city or county still must find a way to meet the task before it.

The reader will recall that we began this volume with idealistic notions of dedication to public service and almost an ideal type public servant, and how difficult that would be to find in practice. In Minot, it is generally agreed that local government did everything it could, under the circumstances, to protect the community. As far the evacuations were concerned, local government understood the risks and told those in that may potentially be flooded to get out while they had time; those orders to evacuate were followed, for the most part. Arguably, there was a trust at work between citizens and government. When the order was given, it was given with due care for protecting the public as much as possible.

Nonetheless, public servants in disasters are on a front line of sorts and require some level of protection. They are, for many practical purposes, engaging in activities that variously cross the politics/administration line, if there is one. As a result, they are called upon to act in a political sphere, where they are not necessarily adept, and where their skills are regardless needed. I found that references to an almost military "chain of command" approach, with subordinate staff "knowing their roles" in protecting the whole, were common.

Recognize for a moment that local officials in Minot had an opportunity to engage in a decision process that is at once overwhelming and solemn. They were asked to make the intractable decisions about what to save in their city. One official commented, "I guess for lack of a better word, within some parts of the community . . . they feel that they were flooded worse than they actually would be because of what the local government did to protect other parts of the city." For its part, government decided to protect city landmarks with a twelve-foot dike that ran north-south just west of North Broadway. Minot State University was also protected by an east-west dike along University Avenue. The choice to protect the needs of many students (4,000) by protecting the college was clear, even though the flood claimed the homes of 116 faculty and staff. Minot State served as an important staging area for relief and volunteer efforts in the response/recovery. An official commented that the Army Corps of Engineers put up the dikes to protect Minot State, but just on the other side of the dike there was a "school that's demolished and the churches, lots of churches, and where you live." These are the sorts of difficult decisions that may confront any disaster or local government professional; they are rare moments to display both courage and leadership when those traits are sorely needed.

Moving on to culture, in one text about North Dakota's neighbor to the northwest, the thesis is offered that Saskatchewan was founded at least partially on myth, and about the belief of the people of the province in a specialness of the place being rooted in that myth.[52] It is not difficult to suggest that, like Saskatchewan, there is a belief in the specialness of North Dakota among those that live there. Weather-wise, like the rest of the Midwestern and Great Plains United States, North Dakota is sometimes a tough place, prone to extremes in heat and cold, severe weather, and howling winds (a FEMA trailer village in a snowy, post-flood Minot is shown in Figure 7.5). In considering the effect of North Dakota on identity, Debra Marquart spoke to the cruelty of the weather that yielded an unusual strength in the state's people.[53] North Dakota is at once spectacularly beautiful and resistant to the entreats of people. Somehow, the people might be more human when disaster strikes, acting with greater empathy and care, than in other places that are not so challenged. Because the land does not care, the people care about each other more. To the extent that the specialness in how the people in Minot see themselves is somehow mythical, it very likely became real in responding to the flood. Kathleen Norris wrote: "here, more demonstrably than in many other places, we need each other to survive."[54]

The theme of community and how people actively care for each other, even if they are not related to one another, was a common element in most of the interviews. It is a core belief. The simple fact that shelters could be opened for flood victims, but that they would go largely unused because of the willingness of people to bring family and even strangers into their homes, is worth mention. Because the Minot area is a valley, not all of the city was devastated or even physically damaged by the flooding. For those areas that

*Figure 7.5*   FEMA housing site, Minot, November 2011, FEMA/Robert Kaufmann.

had been spared, the people in those homes did what most in Minot would apparently do under the circumstances—they helped as much as they could to get their neighbors back on their feet.

The culture of Minot is unique due to the close-knit relationships shared by its citizens. In the Minot case more than the others, the thematic element of community took a central role in describing what had occurred. As an example, the theme of not needing outside help was a point raised again and again. Even though the city recognizes its position as part of a larger intergovernmental structure, and does, in fact, appreciate all the assistance it has received, it identifies with and supports the fierce independent streak of its citizens who do what they can to take care of each other before seeking other aid. This is an important point because it harkens back to the principles of the national response framework—that in ideal situation local governments are on point first and foremost in responding to disaster events before either the states or the federal government get involved. In Minot, that might well have worked. It works because the people have an especially strong work ethic, and are oriented toward moving forward from whatever might happen to them as a matter of existence. There really is no question as to whether they will pick up and move on.

There are several points worth additional discussion in light of culture. The first is the flight of young people from North Dakota. The second is the strong North Dakota work ethic. The third, and perhaps most important, is the role and integration of faith-based organizations in responding to the challenges presented by the hazard event.

A recurrent theme in the interviews is Minot's young people leaving the community and the kind of impact that has. This was echoed in the literature as well as having been a theme: "It is a given that isolated Plains communities cannot hold on to most of the best and brightest who grow up there."[55] One official commented,

> *Prior to oil boom and that prosperity that all hit, it was always said that our greatest export was our youth because there's no jobs . . . not a lot of job opportunities. Not a lot of good paying jobs. Not true anymore so literally we exported a whole generation and they're gone. So when these folks, the elderly, when they flooded there wasn't any family really here to help. Their children are in Georgia or Chicago or wherever. It was really difficult for them. We saw a good number of those folks simply leave and go live with their children. They either sold their house as-is or abandoned them, they just couldn't fix them. Those are the sad, sad cases.*

Not everyone agrees that Minot is a place to escape from; one young staff member working with one of the community partners noted, "I'm a member of the younger generation and a lot of kids my age who want to get out of North Dakota to Minneapolis and Denver, where real life happens. I'm of the opinion that North Dakota is where it's at. I've spent time as a military dependent in other places and I absolutely love North Dakota. I love this community and its people."

The idea of a strong North Dakota work ethic was also a constant theme. Even when danger is threatening an individual, they still want to help others; one official commented that "it's the same old thing that everybody always hears about North Dakota. It's different here than it is across the country. Nobody's standing with the hand out. They're saying we've got to get this done. My daughter, for example, was affected by the flood. They put a dike right on the back of her building and they had to move out. The water was seeping underneath and coming in."

There is little tendency toward reliance on outsiders or even state or federal government. "This is my city. It's not owned by the government. It's not owned by the state. This is my hometown. This is where my grandkids live. This is where family lives and I'm going to keep it up as nice as I can instead of expecting federal government to come in and do it for me." Another official commented similarly: "Midwestern values is that you help yourself, you don't wait for the handout, and you don't wait for somebody to say 'What do they need?' You put your nose to the grindstone and get back at it."

This work ethic was not limited to people helping their families or even just their neighbors. "A lot of it is you just help your neighbor, you know. You know there were stories left and right about complete strangers [helping each other]." People would work tirelessly to get their lives back in order: "Every night after work the guy would go back to his house and do an hour, or an hour and a half, two hours, worth of work trying to get things cleaned up and start the whole process of rebuilding. His weekends were spent there." Another official added, "Nobody is sitting here waiting for people just to fix their ills for them."

To that end, "from a financial standpoint, the economy right now in the United States is what it is, and we've probably gotten every bit of assistance we're going to get. It's up to all of us to do what we can to get our homes back in order." Another official summed it up as "maybe it's faith, that could be what it is. It might be people have learned from their church or their faith to try to take care of their neighbors, and that's important in our everyday life."

Speaking of the role of faith, faith-based organizations played a major role in Minot's response and recovery. Officials agreed that faith was a central part of the whole effort. "I'm always a firm believer in faith base being involved of life in whatever anybody does. Not everybody agrees with me but I have always believed that and not a denomination. Not this denomination does this but a faith base, whatever faith that you were." Where there are volunteers, the "commonality that we have is that their agenda is from a faith-based side."

Hearing officials speak about the role of faith in their response was a moving experience. "The faith-based groups wrap themselves around this community not looking for any credit from any place at all." Another official commented, "What I try to tell campers [volunteers] that are coming in, my goal is to for you to become as blessed as the folks that you are helping. If we do not get so caught up the mechanics that we lose sight of the mission of the ministry, then we have got it made." Indeed, focusing on the mission is more important than individual church differences. This commonality of purpose was echoed in other interviews:

> No one says, "Oh you're from the Lutheran Church, or you're from the Assembly, or you're from the Methodist?" It's "Whose house are we doing today? You need this piece done. Our church does that. You need a lawn plant or we'll take that piece." They're jumping on the long-term recovery bandwagon and volunteering their time to do a piece that they're good at. Nobody cares who gets credit. That's another big piece. A lot of times somebody wants credit. Put my name on it. Put my name on that new building. Nobody cares. They want to get the people back in the homes.

So much of the interviews focused on faith that one could spend a chapter speaking about just this aspect. "From the faith-based side, we are all here to serve. We all speak to the same God. We might have different accents, and it is my role [as an official] to be that interpreter and say, 'Your synod is pushing this piece of this and it is not an opportunity to . . .' Nobody comes here to proselytize. But we certainly share God's love with the people under this banner." The *Summer of Hope* banner of recovery allowed people to show others how their faith has made them strong and gotten them through this trial. The determination to effect so much positive change by connecting these diverse resources shows that, even in the face of suffering, it is possible for public work at the local level to speak to the very best of humanity.

The last part of the original conceptual model is vulnerability. Vulnerability is a real issue in Minot, though in some respects it is less significant than it might be elsewhere because of the level of expertise, concern, and genuine care being brought to bear on those concerns by those in need. There are still a variety of issues that Minot faces that make the community vulnerable in a variety of respects, even though the city and its partners seem positioned to respond favorably to the challenges and appropriately in the event of future hazard events.

First, the low- and moderate-income residents along the river are more vulnerable. One official noted,

> Tale of two cities. Who got flooded? Unlike Bismarck, for example, where the homes along the river are three, four, five hundred, eight hundred thousand dollars, this is probably like most small communities that have a small river running through them. The older homes are by the river which means that those are the homes where retired people live on

*a fixed income. Lower income folks, moderate to low income, that's all they can afford. They're 800 square feet homes . . . Small homes, small lots, but they're affordable houses.*

Many elderly residents live in these homes; this creates a real problem in terms of vulnerability because these residents were already vulnerable due to both their age and the rising cost of living in the city. Elderly residents can be targets for contracting scams. Further, the elderly on fixed incomes do not have the luxury of moving from a flooded house to a trailer to another home of their choosing in Minot; affordable housing is seriously lacking and even if it were available, fixed incomes hamper efforts for the elderly to move on and rise above the flooding.

The portion of the city that was flooded represented a substantial portion of the city's workforce; this represents an impact to not only the individuals, but to the community as a whole. An official commented, "The 20% that was flooded are the workers. They are the service workers. They are the people who provide common, everyday jobs. They are not the managers necessarily, a few in that particular area where the nice homes were, but generally speaking these are all service workers in the valley. Secretaries and service workers in restaurants and cooks and everything else." Because the people on the hills surrounding the valley were not seriously impacted, there is the possibility that they did not understand the extent of the damage felt by the residents along the river. "I could venture to guess that the majority of the North and South Hill folks do not venture down into the valley and see what is going on. They do not want to have to deal with that. Maybe some of our support we get is a way of clearing the conscience."

Where there is obvious shortfall in Minot's response to this flooding event is the problem of flood insurance. The problem is not so much Minot's as it is a national problem, and goes to the heart of a national policy that can entirely miss the point and put great numbers of people at risk. As was noted earlier, in Minot only 10 percent of the population had flood insurance. This is mostly because they felt that they did not need it. Indeed, the common perception of the nature of flooding risk, and that used by official channels, would have indicated to them clearly that floods posed little risk, at least as far as 100-year flooding events are concerned. The problem is that 500-year flooding events are possible, and one has occurred in Minot through a confluence of events. No matter how unlikely it may have seemed to those designing the policy, the policy had unintended consequences for Minot, and put the city at a disadvantage before it even began to respond to the flood. One official commented, "FEMA was saying that after the 1970s and 80s and into the 90s in the flood efforts that were done that we're no longer in a flood plain. That is the other story, I think, that it is important to understand, is that of the 4,152 homes that were damaged in the valley, less than 400 had flood insurance." Another official added,

*We can compare this to Joplin [Missouri tornado event]. In Joplin there were 6,000 homes that were affected, 4,180 here. But everybody is insured for wind damage. [Few homes here] were insured for flood damage. You put that . . . you do not have this insurance policy that is going to pay at least a percentage of your rebuild. Most of these people: there is nothing. It is zero percent. They have gone to FEMA and they have gotten their grants and things like that, but there is no insurance money to help support that.*

Table 7.4 shows a breakdown of interview codes by theme and how these elements impact community resilience.

Vulnerability is shown as an element that decreases resilience, but community aspects and nonprofit and faith-based involvement were elements that had strongly positive impacts. Further, business and economic elements were positioned by interviewees as allowing the community to respond and recover. Federal agency interactions were seen as largely negative, and local government actions, even with communication concerns and occasional

*Table 7.4*   Comparison of themes in interviews by apparent impact on resilience

| Decreasing Resilience | Codes | % of Category | Increasing Resilience | Codes | % of Category |
|---|---|---|---|---|---|
| Vulnerability | 76 | 100.0% | Vulnerability | 0 | 0.0% |
| Federal Agencies/Outside Assistance (Negative) | 58 | 86.6% | Federal Agencies/Outside Assistance (Positive) | 9 | 13.4% |
| Communication/ Skill/Planning | 49 | 50.5% | Communication/ Skill/Planning | 48 | 49.5% |
| Negative Aspects of Local Gov't Response | 9 | 18.4% | Positive Responses by Local Government | 40 | 81.6% |
| Community (negative aspects) | 5 | 6.3% | Community (positive aspects) | 74 | 93.7% |
| Activities of Nonprofits/Faith-based Groups (negative) | 3 | 3.7% | Activities of Nonprofits/Faith-based Groups (positive) | 78 | 96.3% |
| Business/economy (negative) | 2 | 4.8% | Business/economy (positive) | 40 | 95.2% |
| TOTAL | 202 | 41.3% | TOTAL | 289 | 58.7% |
| Loss/Disruption from Flood Event | 54 | | | | |

problems, were seen supportively. Flood loss and disruption was a recurrent discussion point of the interviews, as shown.

## Synthesis and Discussion

Speaking to concerns about the city's ability to deal with flooding, the Minot Daily News ran an article in 2009 that asked, "Are we ready to rumble?"[56] Minot had the distinct advantage of being in a positive position financially to address the problems facing the city and its citizens. But how did it perform in terms of response?

Residents in harm's way took the flood threat seriously; for each of the flood evacuations, people responded to the orders. Citizens listened to official communication, and there was no loss of life as a result of the flood; if city officials had not remained on point with providing information to the public, or if citizens had ignored warnings, an unfortunate hazard event with great loss of property might well have turned out to be something much worse. Minot benefitted from the trust that exists between residents and local government. Local government responded quickly and led with a firmness when it needed to convey the threat that faced the city. When it was realized that communication problems existed, such as the need to control rumors and get information out from a trusted source, it quickly moved to hire a professional PIO.

While it would have been more proactive of the city to have known in advance what to do and how to do it in the face of the flooding threat, there was an adaptability in the response that allowed outcomes for Minot that were relatively favorable, even as so much destruction and suffering presented a tremendous task for all concerned. From a planning and response perspective, Minot had a critically important advantage: the people who needed to be at the table for recovery operations in the city knew and supported each other, from government to nonprofits to industry. There was at least a conceptual understanding of what needed to be done to respond and recover. Like a case with a similar mindset of problem solving among participants that occurred in southern Sweden, for most people involved in the Minot response, "there was no alternative to personal commitment." There was also a "need to break out of existing structures," and the processes involved were all heavily dependent on social capital and free flow of information.[57]

Of the importance of community participation after a hazard event, it was suggested that forums allow "an outlet for people to articulate and solve problems, [which] empowers them to take action, and as a result assists in reducing anxiety and trauma and helps build resilience to cope with future events."[58] It could be said that this is what is occurring in Minot. The city and its partners have provided forums for the public to participate and discuss how the community will move forward and recover. This gives citizens a sense of control and ownership in what will eventually become the city's future. This is beneficial, not only because it helps people feel more comfortable about the planning process as it moves forward, but because the process itself will be more representative of the actual wants and needs of the broader community.

Minot faced the same situation faced by many other communities, not only in the United States but throughout the world, in that improvements to management of a potential hazard likely led the city and the surrounding region into a false sense of security. Improvements to dams and channels in Canada may have led to an assumption that flooding events were being controlled. This is further evidence that governments sometimes do not communicate as well about potential threats as is needed. Citizens, suspecting nothing, sometimes do not buy flood insurance, because they so not see themselves as being at risk from flooding. The result is a mistaken notion of risk.[59]

While disaster communication was noted as a considerable dilemma for local officials, Minot officials understood implicitly the importance of exploring all potential options for funding in supporting the response and recovery. Minot was particularly successful in gaining funding through the Community Development Block Grant Disaster Relief fund—Minot received $112 million through the program as of March 2013. Coupled with other federal funding, the city and region has received more than $632 million toward recovery. In the context of the city receiving the additional funding, Senator John Hoeven remarked that "the people of Minot and Ward County experienced a devastating flood two years ago and though much has been done, the work to recover and rebuild continues," and that the "support will help with housing, business and infrastructure needs beyond those addressed by other forms of public and private assistances to help the region get back on its feet."[60]

Business recovery progressed admirably in Minot after the flood. This may be because Minot has retained the positive social capital that characterizes rural areas; this has positive implications for resilience. It has been proposed that "rural business owners' orientation in times of stability may be toward the community because personal service to a community by a business owner means that a community may loyally patronize and promote a business in return (Besser, 1998). When a natural disaster occurs, however, rural business owners may feel torn between the clear needs of their family or firm and those of the community."[61] Minot's experience tends to reflect Besser's conception of rural business, with community support of local business. Business owners may have had experiences where they were torn between the needs of their firm and those of community; challenges were likely in the response and recovery, but the outcomes for business in the community were quite strong.

With regard to the oil industry's impact, the influx of economic activity has had positive returns for Minot, and this has allowed the community to be much more responsive to the threat from flooding than might have otherwise been possible. The oil industry has been valuable for Minot in this way and has allowed city government to begin to think about infrastructure improvements and planning for future flooding events. In another sense, the industry has caused some of the problems that the city now faces; the city has to address concerns about housing, traffic, and wear-and-tear on the roads and other infrastructure from the crush of people. Crimes of all sorts, and additional students enrolled in public schools, strain systems that have historically had to serve fewer residents. However, the city refuses to be play the victim. Where cities in other parts of the country might respond with requests for outside assistance, Minot has responded with a campaign based on hope as a founding principle. In Minot, we find the "realistic optimism, facing fear, moral compass, religion and spirituality, social support, and resilient role models" of the Southwick and Charney model of resilience,[62] but at a city level. The introduction of a particularly strong work ethic and a sense of uniqueness of place and people make for an even stronger case for community resilience.

The issues surrounding shale oil extraction deserve our attention. A case in point is hydraulic fracturing, or fracking, which has been defined as "the process of pumping, under high pressure, engineered fluids containing chemical and natural additives into the natural gas or oil well. This process creates and holds open fractures in the oil or natural gas, formation . . . [And] allow oil and gas to flow up through the well . . . from previously unavailable sources."[63] Fracking is central to working the Bakken formation, and it might be a bright spot for the business and good for North Dakota, just as it might be a source of groundwater contamination and potentially an area of regulatory concern.

The positive benefits to community and society, including employment, both in the oil industry and indirectly, and reducing dependence on foreign oil, will not be debated here.[64] However, given what occurred with BP in the Gulf of Mexico and the impact of that particular disaster on an already sensitive New Orleans MSA and its Gulf-centered industry[65], the matter of environmental harm, current or future, is worthy of discussion. Just as the BP oil spill had consequences for the Gulf Coast of the United States, with serious potential hazard to human health, quality of seafood, and detrimental effects on tourism to areas where that is the lifeblood of commerce, ignoring environmental quality issues is neither wise for the public good nor responsible business for North Dakota or its companies.

In my view at least, it is not worthwhile to plead exclusively for environmental quality or jobs, when we need both for communities to be sustainable. However, so much money and business is coming into North Dakota all at once, it is worth asking: is this *development*, and how much of what is occurring benefits the whole community? When I visited, to me it felt like Minot and the region were riding a wave of fortune; governing is demanding under these kinds of circumstances, and managing expectations sometimes involves more a coping stance than a proactive planning orientation. The growth challenges an existing way of life—everything from lifestyles of long-term residents to the livelihoods of farmers and ranchers, to the simple serenity and calm of life on the Great Plains.

It was not too long ago that the United States had serious environmental problems due to both a lack of effective regulation and a dearth of enforcement. Hopefully, society as a whole better understands now that there is a price for failure in environmental stewardship. This cost goes beyond simple clean-up, because public health is affected by persistent pollution in ways we do not always fully understand. Articles such as those by Gibson[66] and Karaim[67] suggest a substantial downside to the pollution and rapid development associated with work on the Patch. Schmidt warns of problems seen with fracking associated with shale gas operations, including methane contamination of aquifers from poorly constructed wells.[68] At minimum, the negative aspects should be weighed with the positive, and protection and respect for the environment and residents be made just as much a priority as maintenance of the admittedly beneficial growth that the state has experienced.

There is an opportunity for regulation to work with industry to make the process safer and less environmentally risky. Enforcement of regulations already in place at the state level is vital; inspections not only during construction of wells, but also at disposal and waste sites, are needed to avert groundwater contamination; the state has focused on well construction to considerable benefit, but greater resources for review of disposal and waste sites are vital.[69]

With unemployment being such a great concern, some extreme position such as waiving use of the oil reserves in North Dakota is not preferred. That said, the long-term ramifications of reliance on fossil fuels, including the potential impacts of anthropogenic global warming, are of great consequence and must be considered in the construction of public policy that is most advantageous to the public.[70] This is not an ideological matter—it should be confronted scientifically.

Hill and Olson properly consider North Dakota's "confluence of attributes" as presenting difficult options for balancing nature and the needs of humankind.[71] North Dakota is a beautiful place. The countryside is sprawling, covered with lakes, lush farmland, and rolling pastureland; the gifts of the state are multifaceted. The place seems wide open and enormous. North Dakota is not only a petroleum powerhouse; it is an impressive agricultural producer, a home for wildlife, and even a region showing potential in renewable resources. Global and economic considerations, and the need for self-reliance in the face of pressures from foreign oil, weigh against considerations of social equity, balanced growth, and local/regional environmental protection. Policy reform balances against the desires of market-forces to expand production and use of fossil fuels from the area.[72]

North Dakota exists in the context of market demands and presently sees tremendous compensation for its natural resources, but the Patch has been used before and demand has dropped. Given the contentious nature of fossil fuels in geopolitics, it is possible that the boom in North Dakota will go on for some time, but it difficult to ignore the idea that North Dakota is used by industry, for resources and industry benefit, just as much as it uses industry, for employment and growth. For now, this is perhaps an acceptable trade-off. A theme of interviews conducted for this review was that the oil industry has had positive impacts for Minot, and I would not presume to debate the economic benefits, which are clear and impressive. These benefits have, at least over the short term, led Minot's recovery efforts to the advance of the community as a whole, responding quickly to the 2011 flood without the occasion becoming a media circus. However, Hill and Olson's analysis of the issue raises questions about the future of North Dakota and the assumptions being made.[73] To its credit, Minot has moved to diversify its approaches to economic development and upgrade its infrastructure which bodes well for the city's future. The tight-knittedness of the community is also a point in the city's favor as it moves ahead.

We now move to our final chapter, which evaluates the cases in a comparative light, and proposes some common lessons, extending from the cases themselves, that may contribute to the development of greater resilience in cities and counties.

# 8 Toward Resilient Communities
## Coming to Terms with the Threat

Local governments bear a heavy burden in all that is asked of them in response and recovery efforts, from public and business expectations to infrastructure requirements. This burden is greater when communities have not done the necessary work to improve capability to respond to threats well in advance of the events that may befall them. The work of government to respond to these expectations can be made impossible when communities lack critical capacities. When responding to the needs of businesses, which vary with industry and size, the need for local governments to not only plan but act is obvious. Individuals benefit from competent response and clear communication from strong, steady, and visible leadership at the local level.

We now compare and contrast the case studies, and how these communities have responded to the threats that face them. If they have had to adapt and be more resourceful absent adequate planning, we note those issues. Even in situations where there have been failures, it is crucial that local governments not be dragged down by them—if failures are the greatest teachers, then improvement and innovation in creating more resilient communities may come from the local governments that need them the most.

For cross-case analysis, we follow the general framework offered by Seawright and Gerring, mentioned previously.[1] Table 8.1 summarizes the cross-case comparison.

The New Orleans metropolitan statistical area (MSA) represents an extreme case because of the extreme values of the dependents,[2] namely size of the disaster at the individual business level and apparent vulnerability. We find that the case is indeed extreme for any number of reasons, but the reason for the anomalous outcomes may ultimately lie in the inability of governmental institutions themselves to break through constraints, think proactively and with vision about how to respond to threats, and act as a coordinating force within the community to the extent necessary to fill the leadership void. The historical constraint within institutions is profound. Further, institutional constraint has baffled efforts on the part of well-meaning staff members to push institutions forward, particularly in the case of New Orleans, and their

*Table 8.1* Cross-case comparison

| | New Orleans MSA (Extreme Case) | Palm Beach County (Typical Case) | City of Minot, ND (Deviant/Influential Case) |
|---|---|---|---|
| Relative Size of Event | Unimaginable, even from a safe distance. The event continues to haunt those who suffered through it; the city of New Orleans continues to recover from it. Disaster in every sense of the term. | A severe storm, with surprising damage for the category. Extensive disruption to communities, disproportionate to the size of the storm itself. | Devastating, particularly in light of the vulnerabilities of those directly impacted. Fully 1/4 of the city's residents were directly impacted by the flood, which destroyed not only houses but infrastructure. |
| Institutional Actions | General failure of governance at local, state, and national levels to respond effectively through institutional structures. Public left in desperate scenario. | Public institutions such as procurement and economic development structures worked relatively well in responding to the needs of local businesses. | Local officials worked tirelessly to save their town and did so for a period of months. Lines between institutions and partners blurred-focus was on helping citizens get back on track. |
| Role of Institutional Culture | Highly constrained; political forces frequently clashed with administrative resources to no advance of the organization as a whole; frequent allegations of corruption and individual self-serving behavior; inability to make sense in crisis. In surrounding parishes, less constrained and more supportive of the independent role of administration staff to serve the public interest in fairly carrying out their roles. | Institution supports/encourages action within regulatory limits. Some ability to make sense in crisis which should be further encouraged. Attention to competition in procurement activities and benefit of development to for the community as a whole. | Few constraints noted on partnerships between governments, nonprofits, and faith-based groups in support of response and recovery. Honesty in approach and respect between partners seemed to permeate the conversations. Considerable sensemaking ability-given resources, find the best use to maximize benefit to population in need and consider ways to keep potential problems from turning into disasters. |

*Continued*

*Table 8.1* (Continued)

|  | New Orleans MSA (Extreme Case) | Palm Beach County (Typical Case) | City of Minot, ND (Deviant/Influential Case) |
|---|---|---|---|
| **Vulnerabilities** | Poverty; in certain respects, racial discord, and misunderstanding the value of the rich cultural heritage of New Orleans in its original form in favor of gentrification. Crime. Attention to the special needs of small businesses needed. | Some income disparity. Attention to the special needs of small businesses needed. Housing is expensive and cost of living may be considered prohibitive to some segments of the population, particularly in a hazard scenario. | Vulnerabilities shifting due to the presence of new industry expansions/rapidly growing population. Housing shortages, crime, demands on infrastructure. Home-based businesses have special vulnerabilities. Individuals frequently did not have the flood insurance that they needed; need for clearer communication quickly realized. |
| **General Findings** | Size of disaster on a personal level impacts business recovery. Personal fortunes are closely tied to those of small businesses. Institutional culture has an impact on business recovery. In a larger sense, local government contributes to empowerment of a community, both individuals and businesses, when leadership communicates clearly and has a plan for the way forward. | As expected, size of business on an individual business level affected recovery, which is tied to resilience. Insurance played a role in aftermath capacity. Procurement factor was significant in explaining recovery time. In a larger sense, government in the county was relatively more approachable, communicated well, made better use of its resources with regard to procurement and economic development, and businesses (and individuals) derived benefits from this. However, a larger event may prove a strain on the effectiveness of the system, if it has not been significantly enhanced in the years since Wilma. | According to third-party surveys, businesses returned to normal quickly (within a year). Nearly all businesses came back. This likely has to do with both demand for services, due to the larger economic boom in western North Dakota, and the still strong social capital of the community of Minot. From local government, to nonprofits, to faith-based groups, all worked together on the recovery effort. Vulnerability is a considerable issue for a variety of reasons, but the awareness in the community of the need is accompanied by a generous social safety net. Residents lacked necessary flood insurance. |

| **Institutional Constraint and Sensemaking** | The differing outcomes in New Orleans and surrounding parishes show the impact of constraint or encouragement on government staff-they either fail to make connections to use resources, or they make sense of complex situations and choose well at critical junctures. People were not always honest with each other. Teamwork was lacking. | Encouragement by the institution allows staff members to fully play their roles. This pays off in normal operations but is important in crisis situations as well, when staff must make choices quickly and without all information needed to optimize a choice. Staff members also seem to trust each other and get along well, which bodes well for teamwork at critical times. | Resourcefulness is a matter of culture at the individual level; it permeates institutions throughout the city and region. The work ethic arguably does not allow a lack of resourcefulness or effort, and the individualism, though obvious, does not supplant a clear determination to work as a team in crisis situations. Honesty and trust were relied upon constantly throughout the crisis. |

sensemaking efforts were unable to stem what has become a self-fulfilling prophecy in the face of risk.

The Palm Beach County case was selected because it was thought to be a typical case; the experience with Wilma allows us to explore relationships identified in the disaster literature for vulnerability, extent of disaster, and institutional impacts, on recovery and capacity, without the *clear* interference of other causal paths.[3] Palm Beach County's response rate for the survey was better than that of New Orleans; the environment for accountability was far more open. It is thought that this would be more representative of local government experience in responding to hazards, and perhaps even more of an optimal experience toward which to strive, than that seen in the New Orleans area. The business community expressed greater satisfaction with local government processes than in the extreme case.

Finally, in Minot, we have an influential case, where what was apparently occurring in Minot seemed to challenge preexisting theory and the validity of the conceptual model in its original form.[4] The initial consideration was that the case may be deviant, with a surprising value for time to recovery for such a large portion of the community; the case was therefore worth further exploration. Beyond that, it is apparent that the conceptual model, which considers size of disaster, institutional impacts and culture, and vulnerability, has merit in application in Minot, but the city and its residents have already begun to address matters related to vulnerability in such a way as to blur the lines between institutions, institutional culture, and community, including faith-based groups and nonprofits.

The case is influential in that, if resilience can be learned at all, the best way to learn it might be from watching a close-knit community, with firmly set beliefs, people who trust one another, and a faith beyond itself, and see how its people accomplish response and recovery. We also see that a variety of factors influence community resilience in varying ways. The model of community resilience is intended as a starting point for discussions; exploratory research in this area should eventually lead to a firmer foundation of understanding that will be beneficial to communities. We should not believe that the Minot case is entirely unique—without doubt remarkable communities exist the world over that would respond with similar concern for the whole as did Minot, and a firm foundation for resilience exists in those places. Nonetheless, if social capital, trust, and commitment to community are the "talent," then resilience is merely the refinement of that talent into a skill set that can be ready at a moment's notice.

The remainder of this chapter is presented as lessons emerging from the case studies for communities seeking resilience, between local government institutions, the business community, other private entities, and individuals. Hopefully this approach will encourage discussion of capacity in communities while it inspires all parties at the local level to help make their localities and their regions, stronger and more able to weather whatever comes their way.

*Lesson 1: Work to increase organizational capacity and*
*skill and avoid constraining individual abilities*
Neo-institutionalism holds that institutions influence, and are influenced by, institutional actors. Further, sensemaking provides an added dimension within neo-institutionalism, as individual actors seek to make sense of their position and context within a rapidly changing situation. Actors may be unable to make sense of their context most advantageously—be creative in a manner that benefits the public—if the institution constrains them. Even skillful professionals can be limited by an institution that ignores warning signals and dissuades staff members from responsible action, out of a need to address other organizational priorities. This sort of institutional constraint can turn a hazard event into a disaster. When institutional actors are unable to make sense of a rapidly shifting situation, make adjustments, use resources to greatest advantage, and feel empowered to act on behalf of the institution to do so, the quality of the institution's response will be lower.

This lesson brings this text full circle. The best and brightest of the public sector should, with a mind to fulfill the spirit of the public service, approach their jobs energetically and be supported in such endeavors by the public. Society needs to recognize the role and purpose of government and not only in times of crisis. Intelligence, honesty, and professionalism in public positions, including questioning of role and function when a potential crisis looms, should be rewarded, not serially questioned out of partisan bickering masquerading as concern for fiscal accountability. This societal constraint might even be leading a race to the bottom in terms of what is expected among the public sector itself. Instead of creating a groundswell of service quality, pessimistic societal attitudes may cause public sector employees to become increasingly indifferent, disillusioned, and more likely to look after their own interests at the expense of others. At critical times, this is dangerous and can cost lives.

The public sector should fulfill the role of coordinating response and guiding communities toward resilience. Whether individuals and businesses will play a key role in such efforts, or resist planning for the futures of cities and counties in the face of undeniably heightened risks, and instead portray such activities as wasteful, is the question.

*Lesson 2: Learn about community needs*
One of the greatest lessons a local government might learn, before it needs to respond to a hazard, is what the community actually needs. This may be different than the wants of prominent community groups. Some local governments assume that they know what businesses and individuals need, but may be far from the mark. The best way to find out is to ask—either directly, or by asking organizations that represent groups with functional and access needs, and business groups—and then consider the input in totality.

For example, local governments know that businesses and individuals both need their electric power back on after a disrupting event, but not so many work actively with utility companies to create plans for action in the aftermath of hazard events. Remember that utility companies, while often private entities, nevertheless are providing a good that is essential to modern life and is of great public benefit in allowing a community to recover and find its new normal. Consider also the benefit that can be derived from reaching out to business groups representing minority business owners or businesses owned by women. From a functional needs perspective, contact groups that represent vulnerable residents, to assure that response and recovery efforts will assist individuals in ways that will best benefit all communities within the city or county. All should rise together in response and recovery.

Private interests often need public help as well and may not know how, or who, to ask for it. Utilities were a serious issue in responding to Wilma in south Florida. While power companies can often mobilize armies of repair staff to begin their work, they still may need government help. Security at their locations may be an issue—with theft of generators and supplies being a problem in some places under normal conditions—so it is perhaps necessary that government work to assist this critical industry with additional coverage from law enforcement. This is but one example. New Orleans businesses, that might have been in a position to provide food or lodging for volunteers, were severely hampered after Katrina. One of the opportunities for local government to act positively was through economic development, understanding community needs and working to return power service for business; this also helped general recovery efforts. In Minot, we saw that a lack of places for volunteers to stay impeded recovery efforts, until smart coordination between partners led to the creation of Hope Village. In Palm Beach County, there was recognition that small businesses played a large part in the economic strength of the community, so business recovery centers focused on getting businesses open and running as a critical part of response and recovery after Wilma. This encouraged individuals to start getting back to normal, too.

It is important for communities to support the spiritual needs of citizens. It is apparent that spiritual capacity can help citizens get through events and move on; a lack of capacity in this regard can prevent them from moving forward. Places that invest themselves with this spirituality and partner with organizations that have care of the spirit as their missions may do extremely well in nourishing the needs of the whole population—not only their material needs, but also their need for comfort and compassion in trying times. Spirituality, the closeness of family, and belief in others and self are perhaps just as important as any other aspect of resilience for a community, and should never be overlooked. These lessons were especially obvious in the Minot case study presented here, and in the larger literature on Katrina's effects.

### Lesson 3: Believe in the "new normal"
While the term "new normal" may, like resilience, be increasingly considered passé among researchers in emergency response and recovery, the expression nevertheless implies a reboot to the worldview of a community that can be

helpful. After a hazard event occurs, the world of that individual or business will not be exactly the same as it was before the event occurred. If an individual or business can assume some potential scenarios about what may happen in a hazard event and plan for them, or just become more adaptable to what may occur, they will be ahead of the game. As resilience rewards adaptability, a hazard event may hold opportunity in the new normal, even as the challenges may seem insurmountable over the short term. Employees, customers, and supply chains can move or change in hazard events, so it is beneficial to not only understand, but believe in the "new normal." Adjusting quickly will allow one to protect one's interests in the aftermath.

### Lesson 4: Have a plan and follow it (as far as you can)

Governments, businesses, nonprofits, faith-based groups, and even individuals can perhaps all agree on one lesson—it is always better to have a plan, with clear procedures, and follow it. Plans should not be so detailed that they do not allow for the creativity of the institutional actors. After all, the best attributes of strong, resilient institutions lie in the professionalism, expertise, and adaptability of actors within the institution. Plans should allow public employees to do what they need to do to bring about the best ends possible under any potential circumstance. No amount of planning can speak to all scenarios, but the endless creativity, sensemaking capacity, and ability to work within parameters in a coordinated effort of a well-prepared, educated public sector is the best toolkit we have to be responsive in the face of risks of all kinds. The advice is useful generally. All groups, from families, to businesses, to local government agencies, should have plans and contingencies for how they will respond to known threats.

Creativity is central to truly resilient responses, but procedural consistency has positive outcomes in disasters as well. Plans and procedures reduce fear, to some extent. They allow for some general rules within which actors can operate, and there is an expectation that others will continue to operate by these rules, as well, and bring the totality of the situation back under control. They also outline approaches that have worked in other instances and thus reduce another measure of uncertainty when actors lack experience. Provided that resilience be treated as the circular concept that it is, and that organizational learning constantly reinvests testing of approaches back into the abilities and expertise of the institution, organizations can enhance skills and grow capacity. Skills should exist not only in an institution's actors, but within its procedures and processes so that knowledge can be perpetuated. This brings us to lesson five.

### Lesson 5: Learn from mistakes—your mistakes and those made by others

If an organization cannot learn from its mistakes and history, it will, in true neo-institutional fashion, continue to be constrained by them. Actors within will function as they always have functioned, and will have similar, if not reduced, ability to make sense of situations when they are challenged in the future. Institutions should not only learn from their own mistakes—they should actively learn about mistakes made by other institutions, and follow through on encouraging positive change within to prevent similar errors from occurring. This combination of post-event review and professional development serves to enhance the efficacy of local government response to future events, by encouraging positive official behaviors and removing any perceived institutional constraints. There was variance in the cases as far as ability to engage in this feedback process.

*Lesson 6: Set aside resources for, and engage in,*
*programs that reduce community vulnerability*
At the 2013 Florida Governor's Hurricane Conference, a general call for getting partners to the table went out in one session, along with the admission that it would be tough to actually accomplish. It is trying and expensive to get community partners to work together in advance of a hazard event. Nevertheless, I assert that it is the role of local government to not only schedule such meetings, but to be the charismatic leadership that starts these conversations and keeps them going over the long term. Again, it is not the plan that may develop from these conversations that matters as much as it is the planning process and the relationships that develop from the discussions.

The whole community benefits because there is a shared understanding and belief in the future of the community that may come from response and recovery efforts in the event of a hazard event. There is a belief in the partnership as reality. The relationships are seen as characterized by honesty and a shared future. Local government has the responsibility to present this vision and encourage other officials, from the private and public sectors, to believe in the vision. This may present a unique opportunity for political or administrative leadership at the city or county level, or both. Either way, resilience at the local government level depends on such efforts.

The reason that this is needed is simple to illustrate: consider two hypothetical communities—one that has not sought to reduce existing vulnerability, and one that has. As illustrated in Figure 8.1, optimal community resilience is the same for both communities—not 100 percent resilient, but perhaps as resilient as could be expected with people involved. The upper chart in Figure 8.1 is representative of a community that has relatively less resilience. Existing conditions represent the second bar on each chart, with the difference between the optimal level and existing conditions representing the impact of vulnerability. A hazard event strikes between columns two and three, leaving each community at a "new normal." However, with efforts to eliminate gaps caused by vulnerability through a recursive feedback process, it is possible that each community can grow closer to its optimal level of resilience, reduce vulnerability, and be better prepared for future hazard events. This is an optimistic assessment of the potential to be derived from such feedback, but is within reason.

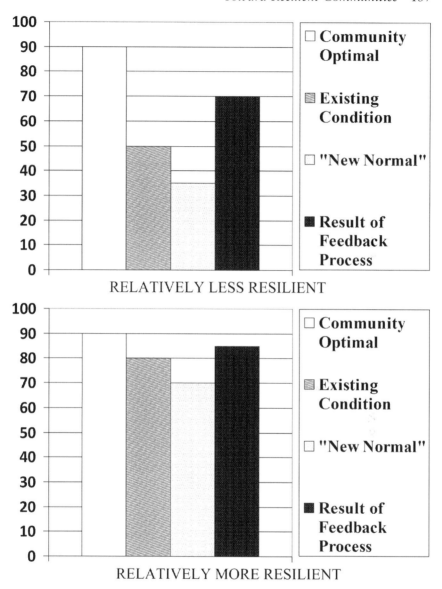

*Figure 8.1*   Optimal community resilience, vulnerability, and the impact of hazard events.

### Lesson 7: Diversified economies respond better to disruption over the long term

All three cases evidence, to greater or lesser extent, the importance of diversifying an economy. We see in New Orleans and Palm Beach County that tourism is a major economic driver. Palm Beach County has made efforts to diversify its economy for many years, and has an advanced economic development skill set; political and administrative forces within government combine with private sector partners to engage the business community to attract, retain, and expand businesses. New Orleans, by its own admission, lacked a push in economic development for the first five years after Katrina. The community still relies heavily on legacy industries—tourism and petrochemical. Diversification of the economies in both cases is worthwhile, but New Orleans has a greater need. Minot, on the other hand, is seeing much of its present growth from the Bakken formation work, and indirect growth is accompanying that push. For a smaller city in North Dakota, challenges and opportunities in responding to these changes abound, but diversification might allow the city to absorb any impacts should a boom-and-bust cycle occur.

Ilan Noy found that, at the national level when controlling for size of disaster, "developing countries face much larger shock to their macro-economies . . . than do developed countries; small economies also seem to be more vulnerable."[5] Generalizing from this, future research might show that small communities with developing economies are more vulnerable to shocks from hazard events than are large communities. Noy also found that institutions matter at the macro level—cost of disaster has a relationship to institutional quality, which was described as either "direct efficiency of the public intervention following the event's onset, or to the indirect impact of an efficient government response in shaping private sector response to the disaster."[6] Clearly government intervention matters in a variety of contexts, at all levels, when it comes to disaster response and recovery.

### Lesson 8: Trust and appreciate the work of government; take care of responders

Because we live in a civil society, citizens have responsibilities, and government work is not simply a job. If we choose to believe in the ideal, and I believe we should, then government employees owe the public their very best, and the public owes itself to try to understand what government does and why, recognizing how institutions are intended to serve the larger public interest. The public sector should more often receive the benefit of the doubt. For those who respond to hazard events, recognize that while crisis events are frustrating, a little patience goes a long way.

### Lesson 9: Communicate

Failure to communicate, within an institution, with public and private sector partners, and perhaps most critically, with the public, can turn a hazard event into a disaster. We saw that with Hurricane Katrina at all levels of government. Capable leadership on the risk of hurricanes, severe weather events, and other threats has kept the public informed about risks, and this has likely limited the potential for catastrophes while raising community resilience. In Minot, communication was found to be a key element of the puzzle of response—so key in fact that the presence of a public information officer was cited as a missing piece of what was otherwise a strong communitywide effort to respond to, and recover from, the flood.

Finally, we consider the community effort of planning for resilience, which closes this chapter and the book.

### *Lesson 10: Resilience is not a one-time workshop—it is an intensive and constant community effort, requiring vision and commitment*

Planning for community resilience is important for reasons beyond the eventual response to a hazard event. To the extent that people in a community hurt or do not have the same level of comfort as others, if local government's involvement can provide access that can alleviate inequality to some extent, this is a positive end and appropriate for local government. Further, too often local governments do not consider the discussion of their futures important because of the constraints of budgets and competing priorities. Not considering the future of cities and counties has resulted in entropy that is both regrettable and upsetting to both residents and the institutions that are intended to serve them. Finally, focusing on the needs in communities can result in benefits now that allow citizens to live fuller lives and contribute to the wealth of the community as a whole. Vulnerabilities and perceptions of lack of productivity are seen as insurmountable in some communities because of a broken-window philosophy; this self-fulfilling prophecy needs to be defeated by a forcefully directed vision on the part of local government that will not take "no" for an answer.

Pursuing resilience is neither easy nor free. Time, energy, and resources need to be accorded to such efforts, which touch everything from the more obvious infrastructure-related issues that put both individuals and businesses at risk, to changing the institutional culture of public entities from an attitude of stasis or entropy to a culture that is forward-thinking, resourceful, and compassionate. Resilient communities place people and partnerships first. Partners in these communities talk to each other not only because someone set a meeting—they want to understand situations from each other's perspectives and genuinely want to support and care for one another. This is about appropriateness in the larger sense; community resilience is a good in itself, and a public obligation.

# Appendix A
## Business Survey

Survey of Businesses on Public Procurement Activities
(Palm Beach County example)

1. How long has your firm been in business? (years)
2. What is your firm's primary industry?

   a) Forestry, Fishing, Hunting, and Agriculture Support
   b) Mining
   c) Utilities
   d) Construction
   e) Manufacturing
   f) Wholesale Trade
   g) Retail Trade
   h) Transportation and Warehousing
   i) Information
   j) Finance and Insurance
   k) Real Estate and Rental and Leasing
   l) Professional, Scientific, and Technical Services
   m) Management of Companies and Enterprises
   n) Administrative and Support and Waste Management and Remediation Services
   o) Educational Services
   p) Health Care and Social Assistance
   q) Arts, Entertainment, and Recreation
   r) Accommodation and Food Services
   s) Other Services (except Public Administration)

3. What is your firm's specialty area of work?
4. What is the race/ethnic group of the primary owner of the firm? If more than one, select as appropriate for the majority owner.

   a) White
   b) Black or African American
   c) Hispanic
   d) American Indian or Alaska Native

    e)   Asian

    f)   Native Hawaiian or other Pacific Islander

    g)   Other

5.  What is the gender of the primary owner of the firm?

    a)   Female

    b)   Male

    c)   Co-owned, Male and Female owners

6.  What percentage of the total value of your firm's work was with the County? (choose from the drop-down list for each period)

## 2002–04

    a)   0%

    b)   0–10%

    c)   10%–24%

    d)   25%–49%

    e)   50%–74%

    f)   75%–100%

## 2005–08

    a)   0%

    b)   0–10%

    c)   10%–24%

    d)   25%–49%

    e)   50%–74%

    f)   75%–100%

7.  What was the average total value of your firm's work with the County for each period?

## 2002–04

    a)   Less than $25,000

    b)   $25,000–49,999

    c)   $50,000–99,999

    d)   $100,000–$149,999

    e)   $150,000–$249,999

    f)   $250,000–499,999

    g)   $500,000–999,999

    h)   $1 million to 2.5 million

    i)   $2.5 million to 5 million

    j)   $5 million to 10 million

  k)  $10 million to 25 million
  l)  $25 million to 50 million
  m)  $50 million to 100 million
  n)  $100 million or more

## 2005–08

  a)  Less than $25,000
  b)  $25,000–49,999
  c)  $50,000–99,999
  d)  $100,000–$149,999
  e)  $150,000–$249,999
  f)  $250,000–499,999
  g)  $500,000–999,999
  h)  $1 million to 2.5 million
  i)  $2.5 million to 5 million
  j)  $5 million to 10 million
  k)  $10 million to 25 million
  l)  $25 million to 50 million
  m)  $50 million to 100 million
  n)  $100 million or more

8.  How many contracts did you have annually with Palm Beach County, on average, during each period?

## 2002–04

  a)  0
  b)  1–3
  c)  4–5
  d)  6–9
  e)  10–15
  f)  16–24
  g)  25 or more

## 2005–08

  a)  0
  b)  1–3
  c)  4–5
  d)  6–9
  e)  10–15
  f)  16–24
  g)  25 or more

9. How many full-time employees did you employ per year, on average, during each period?

## 2002–04

   a) 1–4
   b) 5–9
   c) 10–19
   d) 20–49
   e) 50–99
   f) 100–249
   g) 250–499
   h) 500–999
   i) 1000 or more

## 2005–08

   a) 1–4
   b) 5–9
   c) 10–19
   d) 20–49
   e) 50–99
   f) 100–249
   g) 250–499
   h) 500–999
   i) 1000 or more

10. How many part-time employees did you employ per year, on average, during each period?

## 2002–04

   a) 1–4
   b) 5–9
   c) 10–19
   d) 20–49
   e) 50–99
   f) 100–249
   g) 250–499
   h) 500–999
   i) 1000 or more

## 2005–08

   a) 1–4
   b) 5–9
   c) 10–19
   d) 20–49

e)   50–99
f)   100–249
g)   250–499
h)   500–999
i)   1000 or more

11.   What were your company's average annual gross receipts during each period?

## 2002–04

a)   Less than $25,000
b)   $25,000–49,999
c)   $50,000–99,999
d)   $100,000–$149,999
e)   $150,000–$249,999
f)   $250,000–499,999
g)   $500,000–999,999
h)   $1 million to 2.5 million
i)   $2.5 million to 5 million
j)   $5 million to 10 million
k)   $10 million to 25 million
l)   $25 million to 50 million
m)   $50 million to 100 million
n)   $100 million or more

## 2005–08

a)   Less than $25,000
b)   $25,000–49,999
c)   $50,000–99,999
d)   $100,000–$149,999
e)   $150,000–$249,999
f)   $250,000–499,999
g)   $500,000–999,999
h)   $1 million to 2.5 million
i)   $2.5 million to 5 million
j)   $5 million to 10 million
k)   $10 million to 25 million
l)   $25 million to 50 million
m)   $50 million to 100 million
n)   $100 million or more

12.   How many times were you contacted to solicit your interest for prime contracts or subcontracts for government work related to Hurricane Wilma?

Number of times called for prime contracting opportunities:
Number of times called for subcontracting opportunities:

13. How many contracts or subcontracts did you actually receive from the County related to Hurricane Wilma?

    My firm received County prime contracts related to Hurricane Wilma
    My firm received County subcontracts related to Hurricane Wilma (enter number):

14. How much were the County-funded contracts worth in total?

    Prime Contracts
    Subcontracts

15. If you worked on any federally-funded contracts related to Hurricane Wilma, how much were the federally-funded contracts worth in total?

    Prime Contracts
    Subcontracts

16. Did/does your firm employ local staff (from within Palm Beach County)?

    Yes
    No

17. About what percent of your staff is from Palm Beach County?

18. Rate your level of agreement with the following statements, from "Strongly Agree," to "Strongly Disagree." If you don't know, aren't sure, or the statement is not applicable, indicate "NA"

    a) I have a lot of experience working with this County on its contracts.
    b) When a firm is awarded a contract with this County, it is based more on what they know, than who they know.
    c) I am treated fairly on contracts with this County, by County staff.
    d) I consider my firm to be a partner with the County, helping it to achieve its public goals through my contracts.
    e) My firm has a strong working relationship with elected officials in this County.
    f) The County procurements in which I have been involved have been highly competitive.
    g) My firm works well with small businesses on County contracts.
    h) The County's small businesses are capable and effective.
    i) The actions of lobbyists are a decisive factor in the award of contracts with this County.
    j) I had access to government contracting opportunities that were related to hurricane disaster recovery.
    k) For disaster contracts, the County protected the interests of local businesses in securing disaster recovery work.
    l) During the hurricane recovery process, the award of recovery contracts helped to sustain businesses in my community.

19. Is your firm certified with Palm Beach County or other entities as a small business enterprise, minority/women-owned business enterprise, DBE, etc.?

    Yes
    No

20. If yes, what certifications does your firm hold? Please check all that are applicable.

    a) Minority Business Enterprise
    b) Disadvantaged Business Enterprise (DBE, Federal USDOT)
    c) Women Business Enterprise
    d) Small Business Enterprise
    e) Other

21. If you are a certified business owner, please complete the following questions; all others click "CONTINUE" below.

    a) Certification has helped my firm win prime contracts that it would not have won otherwise.
    b) Certification has helped my firm win sub-contracts that it would not have won otherwise.
    c) Since I became certified, my work on government contracts has increased significantly.
    d) My firm has experienced inconsistencies in County contracting, either being listed on a contract where we did not participate, or experiencing difficulties being paid for work done, and in a timely fashion.
    e) I have had to register an official complaint with the County for a problem I had with a prime contractor.

22. Was your firm located in Palm Beach County/South Florida or doing business in the area at the time of Hurricane Wilma?

    Yes
    No

23. These questions are about your business's recovery experience after Hurricane Wilma. Your comments and information are invaluable to our better understanding how businesses in general dealt with the hurricane and its impacts.

    a) How many days was your business closed due to Hurricane Wilma?
    b) How much damage did your business suffer financially? (In dollars)
    c) How much damage did you suffer personally, in terms of financial impact?

24. Did the personal damages you sustained adversely impact your business recovery efforts?

    Yes
    No

25. In the immediate aftermath of Hurricane Wilma, what capacity did your business have to perform work, compared with your normal business capacity (normal being 100%)?

    a) 100%
    b) 90%
    c) 80%
    d) 70%
    e) 60%
    f) 50%
    g) 40%
    h) 30%
    i) 20%
    j) 10%
    k) 0%

26. Did your business's capacity return to normal?

    Yes
    No

27. If your firm has returned to normal, work-wise, how many months did it take to return to normal?

28. What is your firm's current capacity, with full capacity being 100%?

    a) 100%
    b) 90%
    c) 80%
    d) 70%
    e) 60%
    f) 50%
    g) 40%
    h) 30%
    i) 20%
    j) 10%
    k) 0%

29. Did your business apply for any financial assistance from governmental sources (federal, state, local) to assist you with recovery? For example, a loan from the Small Business Administration?

    Yes
    No

30. If you received assistance, what kind of assistance did you receive? State the source of assistance and the amount you received.

31. How would you describe the application process for assistance?

    a) Very Easy.
    b) Easy.

c) Neither Easy Nor Difficult.
d) Difficult.
e) Very difficult.
f) Other

32. Was your business insured?

Yes
No

33. Did your business have flood insurance at the time of the hurricane?

Yes
No

34. Did you have to relocate temporarily or permanently?

a) Temporarily.
b) Permanently.
c) My firm did not relocate.

35. Do you have any comments on your business experience with working with Palm Beach County on its contracts, the opportunities available to you in government procurement, or your experience with recovering from the hurricane? Please use this space to share your comments.

# Appendix B
## Interview Questions

## 1. NEW ORLEANS MSA & SOUTH FLORIDA INTERVIEWS (2010)

The following represents the general form of the interview questionnaire. Some questions were adjusted slightly to reflect the roles of the interviewee and maintain relevance to his or her specific work. When individual questions were not relevant to an official's role, these questions were not asked. However, all interviews used questions that included the basic concepts highlighted in these examples.

1. Tell me a little about contracting with (name of entity)—solicitations, contracts, etc. What is (name of agency)'s role in assuring fairness?
2. If I am a firm interested in contracting with (entity), what would be your advice for success?
3. How would you characterize your local procurement system? Is it efficient, difficult to work with, or somewhere in between? What are some good and not so good characteristics?
4. Do you see a concern with undue influence in the award of (entity) contracts? Has that been an issue in the past?
5. What is (entity) doing to support business retention and expansion? Is this an area that is a priority?
6. What do you think about the local small business program? Does it serve the purpose(s) it was intended to serve?
7. What are your priorities for economic development? Are there targeted industries that have interest for you?
8. How has disaster recovery proceeded for (entity)? What has worked well? What has worked less well?
9. When disaster struck your community, contracting decisions were made with respect to recovery. Were those decisions basically made the same as contracting decisions are normally made? If they were different, how were they different?
10. Do you find that the same prime contractor/subcontractor teams receive most contracts of a certain type?

11.  How does (entity) foster business diversity?
12.  In (entity), to be successful in business, is it more about what you know, or who you know, or a combination of both?
13.  How has the business community in (entity) responded to emergencies, like the hurricane?
14.  How has government and business worked together to respond to and recover from challenges like hurricanes? What has worked well?
15.  What makes (entity) different from a business perspective?

## 2.  MINOT INTERVIEW QUESTIONS

### A.  *Initial Phone Interview Questions (January–February 2012)*

1.  Did local government perceive the risk of flooding in Minot? To the extent of what actually occurred?
2.  How would you characterize the local government's response to the flooding?
3.  How would you characterize local government's partnership with the community in Minot? How did this help with the initial response and recovery efforts?
4.  What are the strongest community organizations in Minot? How did these groups participate in recovery planning? How about the smaller groups? How were they involved?
5.  Have any organizations declined to participate in your recovery planning?
6.  How is Minot reaching out to assist vulnerable groups within the community?
7.  Moving forward, what are the biggest challenges facing Minot after the flooding? How do you see community partnership forged in the flooding helping overcome the city to overcome those challenges?

### B.  *Government Questions (July, 2012)*

1.  Tell me about government's role in responding to the flood.
2.  Was the response effective? Efficient?
3.  What could have been done better?
4.  How would you characterize the Minot community? What makes it different or special?
5.  How is government trying to be responsive to the needs of the business community in disaster?

### C.  *Finance/Procurement Questions (July 2012)*

1.  Tell me about government's role in responding to the flood.
2.  Was the response effective? Efficient? How so?
3.  What could have been done better?

4. What makes Minot different or special? How would you characterize Minot?
5. How is government trying to be responsive to the needs of the business community in disaster?
6. Does Minot use e-procurement? What happened with procurement during/ after the flood?
7. Are there concerns with fairness in procurement? How does Minot assure procurement fairness?
8. How do you assure accountability and transparency during an emergency/ disaster? Would you do anything differently?

## D.   *Business Group Officials (July 2012)*

1. Does government work well with/understand economic development? If yes, how so?
2. Is government an effective partner in responding to disaster needs for businesses?
3. How would you characterize the business communities response to the flood?
4. How has recovery gone?
5. What would you do differently to encourage business resilience?
6. What is your advice for other communities facing disaster?

## E.   *Nonprofit Community Agencies/Partner Groups (July 2012)*

1. Tell me about your organization and its role in the flood response.
2. How would you characterize the Minot community?
3. How did local government respond? Was the response effective? Efficient?
4. What could have been done better?
5. What makes Minot different or special? What about the people here?
6. How is Minot serving vulnerable populations with special needs?
7. What can you tell me about how the community interacts? Business/ government/nonprofits/citizens?
8. What are the long-term plans for flood response in Minot, to provide for a more resilient community?
9. Anything else you'd like to mention?

# Notes

**NOTES TO THE PREFACE**

1. Jackson, "Responsibility Versus Accountability," 70.
2. Stivers, "So Poor," 54.
3. Grazulis, *Significant Tornadoes*, 1177–8.
4. Swan, *Canton*, 92.


**NOTES TO CHAPTER 1**

1. Paton and Johnston, *Disaster Resilience.*
2. Colten, Kates, and Laska, "Community Resilience," 2.
3. Ibid; National Research Council, *"Building Community Disaster Resilience."*
4. Kates, Colten, Laska, and Leatherman, "Reconstruction of New Orleans."
5. Olshansky and Johnson, *Clear as Mud.*
6. Tierney, "Businesses and Disasters."
7. Tierney, Lindell, and Perry, *Facing the Unexpected.*
8. Ibid, 9.
9. Munich RE, 2012.
10. Munich RE, 2013.
11. Schneider, *Disaster,* 5.
12. Reich, *Aftershock.*
13. Olshansky and Johnson.
14. Davis, "Gentrifying Disaster."
15. Paton and Johnston.
16. Esnard, Sapat, and Mitsova, "An Index of Relative Displacement Risk . . ."
17. Seawright and Gerring, "Case Selection Techniques."


**NOTES TO CHAPTER 2**

1. Holling, "Resilience and Stability," 14.
2. Ibid, 21.
3. Brand and Jax, "Focusing the Meaning(s) of Resilience."
4. Janssen and Ostrom, "Resilience, Vulnerability, and Adaptation," 237.
5. Gallopín, "Linkages Between Vulnerability, Resilience, and Adaptive Capacity," 301–201.
6. Folke, "Resilience," 263.
7. Ostrom, "A Diagnostic Approach," 15181.

8. Adger, "Social and Ecological Resilience," 347.
9. Southwick and Charney, *Resilience,* 6.
10. Mitroff, *Why Some Companies Emerge Stronger . . .*
11. Sapat, "Multiple Dimensions of Resilience," 74.
12. Elran, "Measuring Societal Resilience," 52.
13. Elran, 49.
14. Elran, 54–55.
15. Fink, *Crisis Management.*
16. Zolli and Healy, *Resilience,* 237.
17. Elran, 55.
18. Browning, *Men and Women and Sordello,* 187.
19. Schneider, 6.
20. Schumacher and Strobl, "Economic Development and Losses Due to Natural Disasters."
21. Gilbert, "Studying Disaster," 15.
22. Horlick-Jones, "Modern Disasters as Outrage and Betrayal," 312.
23. Hewitt, "Excluded Perspectives in the Social Construction of Disaster," 332.
24. Paton and Johnston 2006, 97–98.
25. Ibid, 98.
26. Ibid, 99.
27. Alesch, Arendt, and Holly, *Managing for Long-Term Community Recovery in the Aftermath of Disaster,* 59–61.
28. Ibid, 54.
29. Ibid, 56–7.
30. Ibid, 108–9.
31. Rubin, Saperstein, and Barbee, *Community Recovery from a Major Natural Disaster.*
32. Berke, Kartez, and Wenger, "Recovery after Disaster."
33. Richardson, Gordon and Moore, *The Economic Impacts of Terrorist Attacks.*
34. Rose, "Defining and Measuring Economic Resilience to Disasters."
35. Mitchell, *Crucibles,* 12.
36. Solecki, "Environmental Hazards and Interest Group Coalitions," 454.
37. Ibid.
38. Bankoff, "The Historical Geography of Disaster," 29.
39. Alesch, Arendt, and Holly, *Managing for Long-Term Community Recovery.*
40. Watkins, "Economic Institutions under Disaster Situations," 480–81.
41. Sclar, *You Don't Always Get What You Pay For.*
42. Roland, *Privatization,* xii.
43. Cutter, *Living with Risk.*
44. Solecki.
45. Solecki.
46. Arakida, "Measuring Vulnerability."
47. Bankoff, 30.
48. Vatsa, "Risk, Vulnerability, and Asset-Based Approach to Disaster Risk Management," 28.
49. Barnshaw and Trainor, "Race, Class, and Capital," 93.
50. Bankoff.
51. Tierney, Lindell, and Perry, *Facing the Unexpected.*
52. Ibid, 220.
53. Tierney, "Businesses and Disasters."
54. Ibid, 277–278.
55. Ibid, 281.
56. Light and Gold, *Ethnic economies.*
57. Fothergill, "The Neglect of Gender in Disaster Work," 22.

58. Enarson and Morrow, "Women Will Rebuild Miami," 194.
59. Quoted in Tierney, Lindell, and Perry, *Facing the Unexpected*, 179.
60. Dash, Peacock, and Morrow, 1997, 217.
61. Tierney, Lindell, and Perry, *Facing the Unexpected*, 180.
62. Ibid, 181.
63. Bea, *Federal Disaster Policies After Terrorists Strike*, 110.
64. See St. Petersburg College Center for Public Safety Innovation, National Terrorism Preparedness Institute's *Emergency Response for People Who Have Access and Functional Needs: A Guide for First Responders* (2011), for more information about responding to individual vulnerabilities. http://terrorism.spcollege.edu/SPAWARAFN/index.html
65. Ibid, 160.
66. Hobbes, *Leviathan*; Locke, *Two Treatises of Government*; Montesquieu and Richter, *The Political Theory of Montesquieu*.
67. Stinchcombe, "On the Virtues of the Old Institutionalism."
68. Peters, *Institutional Theory in Political Science*, 6–10.
69. North, *Institutions, Institutional Change, and Economic Performance*, 19.
70. Simon, "Rationality in Psychology and Economics."
71. Denhardt and Denhardt, "The New Public Service," 554.
72. Hayek, *The Constitution Of Liberty*, 23.
73. Niskanen, *Bureaucracy and Representative Government*, 5.
74. Whyte, *Organization Man*.
75. Schilder, *Government Failures and Institutions in Public Policy Evaluation*, 18.
76. Ibid.
77. Weick, *The Social Psychology of Organizing*, 5.
78. Ingram and Clay, "The Choice-within-Constraints," 527.
79. Hayek, 24.
80. Meyer and Rowan, "Institutionalized Organizations"; Zucker, "The Role of Institutionalization in Cultural Persistence."
81. Meyer and Rowan, "The Structure of Educational Organizations."
82. Greenwood, Oliver, Suddaby, and Sahlin, "Introduction." *The SAGE Handbook of Organizational Institutionalism*, 4.
83. Ibid, 8.
84. Ibid, 16.
85. Ibid, 17.
86. Ibid, 20.
87. Ingram and Clay, 525.
88. Peters, 19.
89. Miller, "The Norm of Self-Interest."
90. March and Olsen, "Elaborating the New Institutionalism," 159.
91. Mantzavinos, *Individuals, Institutions, and Markets*, 95.
92. North, *Institutions, Institutional Change, and Economic Performance*, 7.
93. Ibid, 5.
94. Ibid, 3.
95. Clingermayer and Feiock, *Institutional Constraints and Policy Choice*.
96. Ibid.
97. Kolb, *Encyclopedia of Business Ethics*, 414–5.
98. North, *Institutions, Institutional Change, and Economic Performance*.
99. March and Olsen, "Elaborating the New Institutionalism," 489.
100. DiMaggio and Powell, "The Iron Cage Revisited."
101. March and Olsen, "The New Institutionalism," 744.
102. Peters, *Institutional Theory in Political Science*, 19.
103. Suleiman, *Dismantling Democratic States*, 52.
104. March and Olsen, "Elaborating the New Institutionalism," 164.

105. Deephouse and Suchman, "Legitimacy in Organizational Institutionalism," 50.
106. Ibid, 51 and 57.
107. Leicht and Fennell, "Institutionalism and the Professions," 431.
108. Greenwood, Oliver, Suddaby, and Sahlin, "Introduction," 29.
109. Weick, *The Social Psychology of Organizing*.
110. Gephart, "Crisis Sensemaking and the Public Inquiry"; Powell and Colyvas, "Microfoundations of Institutional Theory," 283.
111. Weber and Glynn, "Making Sense with Institutions."
112. Weick, *The Social Psychology of Organizing*, 86–7.
113. Weick, "The Collapse of Sensemaking in Organizations."
114. Johannisson and Olaison, "The Moment of Truth," 58.
115. Weick, *Sensemaking in Organizations*, 93.

## NOTES TO CHAPTER 3

1. Hewitt, "Excluded Perspectives," 333.
2. Kunreuther, "Risk and Reaction," 42.
3. Esnard, Sapat, and Mitsova, "An Index of Relative Displacement Risk to Hurricanes." See also Xiao and Van Zandt, "Building Community Resiliency."
4. Ruchelman, "Natural Hazard Mitigation and Development," 56.
5. "Planning Building and Sustaining Public/Private Partnerships," Lecture notes (FEMA), 2013 Governor's Hurricane Conference, Fort Lauderdale, FL, May 6, 2013.
6. Rosen, *Public Finance*, 53.
7. Musgrave, "A Multiple Theory of Budget Determination," 341.
8. Smith, *Wealth of Nations*.
9. Pigou, *Wealth and Welfare*, 157.
10. Orbach, *Regulation*, 285.
11. Ibid, 286.
12. Sims, *Managing Organizational Behavior*, 7.
13. Woods, Dekker, Cook, Johannesen, and Sarter, *Behind Human Error*, 46.
14. Ibid, 49.
15. Donahue and O'Keefe, "Universal Lessons from Unique Events," 79.
16. Woods et al., *Behind Human Error*, 49–50
17. Perrow, *Normal Accidents*, 359.
18. Ibid, 359–360.
19. Taleb, *Fooled by Randomness*; Taleb, *The Black Swan*.
20. Handmer and Dovers, *Handbook of Disaster and Emergency Policies*.
21. Gerstein and Ellsberg, *Flirting with Disaster*, 163.
22. Halter, Arruda, and Halter, "Transparency to Reduce Corruption?"
23. Janis, *Groupthink*.
24. Miller and Rivera, *Community Disaster Recovery and Resiliency*, xl.
25. Senge, "The 'New Institutionalism' in Organization Theory," 81.
26. Wry, "Does Business and Society Scholarship Matter to Society?" 151.
27. Zolli and Healy, *Resilience*, 274.
28. Elhauge, "Does Interest Group Theory Justify More Intrusive Judicial Review," 32.
29. Kettl and Fesler, *The Politics of the Administrative Process*.
30. Harris, *Political Corruption*, 45.
31. Clawson, Neustadtl, and Weller. *Dollars and Votes*, 29.
32. Ibid, 30.
33. Alesch, Arendt, and Holly, *Managing for Long-Term Community Recovery*.

34. Planning guides such *Post-Disaster Redevelopment Planning: A Guide for Florida Communities,* produced by the Florida Department of Community Affairs and the Florida Division of Emergency Management, provide a wealth of ideas about how to engage in post-disaster planning. As of this writing, the guidebook is available at: http://www.floridajobs.org/fdcp/dcp/PDRP/Files/PDRPGuide.pdf.
35. Weick, *The Social Psychology of Organizing,* 10–11.
36. Atkinson and Bridge, *Gentrification,* 2.
37. Krishna, *Globalization and Postcolonialism,* 121.
38. Atkinson and Bridge, *Gentrification,* 2.
39. Fincher and Jacobs, *Cities of Difference,* 195.
40. Handal, "Rebuilding City Identity through History," 53.
41. Anderson, *Imagined Communities.*
42. Ibid, 5.
43. Zukin, *Landscapes of Power.*
44. Boyer, "Cities for Sale," 204.
45. Lipman, "The Cultural Politics of Mixed-Income Schools and Housing," 216.
46. Robertson, "Downtown Redevelopment Strategies in the United States," 429.
47. Blomley, *Unsettling the City,* 52.
48. Runyan, "Small Business in the Face of Crisis," 20.
49. De Mel, McKenzie and Woodruff, "Enterprise Recovery Following Natural Disasters."
50. Irvine and Anderson, "Small Tourist Firms in Rural Areas," 234.
51. LeSage, Pace, Campanella, Lam, and Liu, "Do What the Neighbours Do," 163.
52. Altay and Ramirez, "Impact of Disasters on Firms in Different Sectors," 61.
53. Webb, Tierney, and Dahlhamer, "Predicting Long-Term Business Recovery from Disaster," 46–7.

## NOTES TO CHAPTER 4

1. Qiao, Thai, and Cummings, "State and Local Procurement Preferences," 379.
2. Prier, McCue and Bevis, "Making It Happen," 652.
3. Thai, "Public Procurement Re-Examined," 10.
4. Ibid.
5. Ibid, 11.
6. Qiao, Thai, and Cummings, "State and Local Procurement Preferences," 396.
7. Macfarlan, "From Philosophy to Practice."
8. *Code of Federal Regulations,* Federal Acquisition Regulations System, title 48, sec. 1.102–1(b).
9. Wilson, *Bureaucracy,* 126.
10. Kettl and Fesler, *Politics of the Administrative Process,* 313.
11. Thai, "Public Procurement Re-Examined," 13.
12. Kettl and Fesler, *Politics of the Administrative Process,* 314.
13. Ibid, 315–6.
14. Ibid, 317–9
15. Miller and Fox, *Postmodern Public Administration,* 36.
16. Constantino, Dotoli, Falagario, and Fanti, "Using Fuzzy Decision Making for Supplier Selection in Public Procurement," 404.
17. Ibid.
18. Kettl and Fesler, *Politics of the Administrative Process,* 320.
19. Ibid.
20. Martin, *Federal Statements of Work,* 590.

21. *Code of Federal Regulations,* Federal Acquisition Regulations System, title 48, sec. 3.104–7.
22. Kettl and Fesler, *Politics of the Administrative Process,* 320.
23. Ibid.
24. Kovács, *Enhancing Procurement Practices,* 177–178.
25. Zullo, "Public-Private Contracting and Political Reciprocity."
26. Miller, *Principles of Social Justice,* 93.
27. Zajac, *Political Economy of Fairness,* 141.
28. *Code of Federal Regulations,* Federal Acquisition Regulations System, title 48, sec. 6.302.
29. Thai, "Public Procurement Re-Examined," 7.
30. Waugh, "Removing The Obstacles To E-Procurement Adoption."
31. Dallimore, "The Ten Commandments of Ethical Government Purchasing."
32. O'Looney, *Outsourcing State and Local Government Services,* 158–9.
33. Adams and Catron, "Communitarianism, Vickers, and Revisioning American Public Administration," 53
34. Hewitt, "Excluded Perspectives," 322.
35. Atkinson and Sapat, "After Katrina."
36. Oberle, Stowers, and Darby, "A Definition of Development."
37. Blanco, "Preliminary Assessment of Statutory Compliance . . ."
38. Ibid, 4.
39. Partridge and Rickman, "Do We Know Economic Development When We See It?," 35.
40. Ibid, 18.
41. Clingermayer and Feiock, *Institutional Constraints and Policy Choice.*
42. Buttress and Macke, "Energizing Entrepreneurs," 20.
43. Clingermayer and Feiock, *Institutional Constraints and Policy Choice.*
44. Elkin, "State and Market in City Politics."
45. Nafziger, *Economic Development,* 131.
46. Hawkins and Feiock, "Joint Ventures, Economic Development Policy, and the Role of Local Governing Institutions," 333.
47. Lynn, "Succession Management Strategies . . ."
48. Fleischman and Green, "Organizing Local Agencies to Promote Economic Development."
49. Saiz, "Politics and Economic Development."
50. Long, "Overcoming Neoliberal Hegemony in Community Development."
51. Jennings, *Business,* 169.
52. Brown, Hayes, and Taylor, "State and Local Policy, Factor Markets, and Regional Growth," 41.
53. Cortés and Rizzello, "Hayek's Theory of Knowledge and Behavioural Finance."
54. The International Economic Development Council has provided, through the sponsorship of the U.S. Economic Development Administration, an excellent resource for economic development recovery through the *Restore Your Economy* website. The site includes information not only on roles played by economic developers but additional case studies, reports and slide presentations. See http://restoreyoureconomy.org/.
55. Preuss, "On the Contribution of Public Procurement to Entrepreneurship and Small Business Policy."
56. Haltiwanger and Krizan, "Small Business and Job Creation in the United States."
57. Haltiwanger, Jarmin, and Miranda, "Who Creates Jobs?"
58. Landström, *Pioneers in Entrepreneurship and Small Business Research,* 183.
59. Blackburn and Schaper, *Government SMEs and Entrepreneurship Development,* 1.
60. Woods and Muske, "Economic Development Via Understanding and Growing a Community's Microbusiness Segment," 191.

61. Noon, "The Use of Racial Preferences in Public Procurement for Social Stability," 611.
62. Ibid.
63. Atkinson, "Small Business Development."
64. Brandeis, *Business*, 2.
65. Prier, McCue, and Behara, "The Value of Certification in Public Procurement," 513.
66. Story, "As Companies Seek Tax Deals, Governments Pay High Price."
67. Quoted by Smith, in "Competition and Transparency," 85.
68. Heisler, "A Theoretical Comparison . . ."
69. See the *New York Times* page on incentives, including the article series and data, at: http://www.nytimes.com/interactive/2012/12/01/us/government-incentives.html.
70. Prier, McCue, and Behara, "The Value of Certification in Public Procurement."
71. Atkinson and Sapat, "After Katrina."

## NOTES TO CHAPTER 5

1. National Weather Service New Orleans, Louisiana, "Urgent Weather Message, WWUS74 KLIX 281550," August 28, 2005, http://www.nws.noaa.gov/os/ctamarker/resources/NPWLIX-cta.txt.
2. National Hurricane Center, "2005 Atlantic Hurricane Season."
3. Tidwell, *The Ravaging Tide*, 92.
4. Federal Emergency Management Agency, "By the Numbers."
5. Levitt and Whitaker, *Hurricane Katrina*, 2.
6. Knabb, Rhome, and Brown, "Tropical Cyclone Report, Hurricane Katrina."
7. Picou and Marshall, "Katrina as Paradigm Shift," 14.
8. Keim and Muller, *Hurricanes of the Gulf of Mexico*, 29.
9. Farber and Chen, *Disasters and the Law*, 4.
10. Keim and Muller, *Hurricanes of the Gulf of Mexico*, 17; Picou and Marshall, "Katrina as Paradigm Shift," 14.
11. Farber and Chen, *Disasters and the Law*, 5.
12. Ibid, 6.
13. Picou and Marshall, "Katrina as Paradigm Shift," 11.
14. Knabb, Rhome, and Brown, "Tropical Cyclone Report, Hurricane Katrina," 13.
15. Blake, Landsea, and Gibney, "The Deadliest, Costliest, and Most Intense United States Tropical Cyclones . . ."
16. Tidwell, *The Ravaging Tide*, 87.
17. Picou and Marshall, "Katrina as Paradigm Shift," 2.
18. Davis, *A Failure of Initiative*.
19. Faust and Carlson, "Devastation in the Aftermath of Hurricane Katrina as a State Crime."
20. U.S. Census Bureau, Orleans Parish, Louisiana QuickFacts, http://quickfacts.census.gov/qfd/states/22/22071.html.
21. Frailing and Harper, "Crime and Hurricanes in New Orleans," 58.
22. Ibid.
23. Ibid, 61.
24. Ibid, 63.
25. Whelan, "New Orleans," 277.
26. Greater New Orleans, Inc. Regional Economic Alliance, Mission & Vision, http://gnoinc.org/about-us/mission-vision/, 2013.
27. Edwards, "Obstacles to Disaster Recovery."

28. Briseno, "A Chronicle . . ."
29. Greater New Orleans, Inc., Industry Sectors, http://gnoinc.org/industry-sectors/, 2013.
30. Perrow, *The Next Catastrophe.*
31. Bill Frogameni, "Florida on South Florida: Housing Bust 'Created Opportunities'," South Florida Business Journal, December 10, 2010.
32. Slaymaker, "A New Plan B."
33. Frailing and Harper, "Crime and Hurricanes in New Orleans," 63.
34. Martin, "New Orleans 'Shell Shocked' Documentary . . ."
35. "Mother's Day Second-Line Shooting . . ."
36. Martin, "After 'Tragic' Mother's Day Mass Shooting . . ."
37. Ward and Le Coz, "Courting the Creative Class."
38. Simerman, "In Briana Allen Killing . . ."
39. Barnshaw and Trainor, "Race, Class, and Capital," 94.
40. Leong, Airriess, Li, Chen, and Keith, "Resilient History and the Rebuilding of a Community."
41. Alesch, Arendt, and Holly, *Managing for Long-Term Community Recovery.*
42. Crutcher, *Tremé.*
43. Lewis, *New Orleans.*
44. Creswell, *Research Design,* 146.
45. Denzin and Lincoln, *Handbook of Qualitative Research,* 532.
46. Callegaro and Disogra, "Computing Response Metrics for Online Panels."
47. Bureau of Governmental Research, "Contracting with Confidence . . ."
48. Bureau of Governmental Research, "Private Services in the Public Interest."
49. Additional methodological and analytical details for this and the Palm Beach County case are available in my dissertation, which is freely available online in pdf format. See Atkinson, *An Evaluation of the Impact of Local Government Institutions on Business Resilience in Disaster.*
50. For OLS regression in this exploratory approach, there is enough continuousness in the variable to avoid strong bias in the results. Multinomial logistic regression could be used, but there is potential to lose information in the analysis. See Owuor, "Implications of Using Likert Data in Multiple Regression Analysis," and Starkweather and Moske, "Multinomial Logistic Regression."
51. Atkinson, *An Evaluation of the Impact . . .*
52. Clandinin et al., *Composing Diverse Identities,* 32.
53. McNabb, *Research Methods in Public and Nonprofit Management,* 302.
54. Ibid, 303.
55. Ibid, 310.
56. Whelan, "New Orleans."
57. Liu and Vanderleeuw, *Race Rules*; Wilkie, "Politics."
58. Runyan, "Small Business in the Face of Crisis," 17.
59. Taylor and Brown, "Illusion and Well-Being."
60. Runyan, "Small Business in the Face of Crisis," 19
61. Ibid, 21.

## NOTES TO CHAPTER 6

1. U.S. Census Bureau, "Palm Beach County, Florida QuickFacts," *http://quickfacts.census.gov/qfd/states/12/12099.html.*
2. Mooney, *Storm World,* 177.
3. Pasch, Blake, Cobb, and Roberts, "Tropical Cyclone Report, Hurricane Wilma, 15–25 October 2005."
4. Marcus and Kestin, "Deadliest Hurricane Category."

5. Wood, "Investigators Find Wilma Damage Passes Expectations."
6. Blake, Landsea, and Gibney, "The Deadliest, Costliest, and Most Intense United States Tropical Cyclones . . ."
7. Pasch, Blake, Cobb, and Roberts, "Tropical Cyclone Report, Hurricane Wilma, 15–25 October 2005."
8. Mayk and Zollo, "Wilma's Destruction Baffles FPL Officials."
9. Pasch, Blake, Cobb, and Roberts, "Tropical Cyclone Report, Hurricane Wilma, 15–25 October 2005," 5.
10. Florida Power & Light (FPL), "FPL's First Full Day of Restoration Gets Underway."
11. Florida Power & Light (FPL), "Remaining 14 Percent without Power Post-Wilma."
12. Schultz and Sorentrue, "Depression-Level Unemployment Dampens Spirits . . ."
13. Gupta and Kleinberg, "Hardest-Hit Cities Get Handle on Cost of Wilma's Damage."
14. Pensa, "No Signals, Debris Make Driving Risky."
15. Smith, "Affordable Housing Proposals Grab Delray."
16. Huriash, "Wilma Wound Still Hurts."
17. Douthat and Musgrave, "Officials from State to Local Levels Join Forces."
18. Ibid.
19. Prohaska, "Residents Start Prepping for Wilma."
20. Pounds and Horvath, "Storm Weariness, Wariness Evident."
21. "Chiefs, Dolphins Get Jump on Wilma."
22. "Wilma's Impact on the Area."
23. Anderson, "Wilma's Wrath."
24. Associated Press, "Gov. Bush Criticizes State's Storm Effort."
25. Ibid.
26. "Citibank Reopens Bank Branches . . ."
27. "Beacon Council Conducts Local Business Needs and Damage Assessment Survey."
28. Reppen and Openshaw, "Governor Bush Announces Emergency 'Bridge Loans.'"
29. Boodhoo, "Forget Wilma and Focus on Area's Strengths."
30. Ibid.
31. Ibid.
32. Please see Atkinson, *An Evaluation of the Impact of Local Government Institutions . . .* for detail, including descriptive statistics, collinearity tables, response population information, and OLS regression analysis detail for the Palm Beach County and School Board of Palm Beach County case.
33. Childs and Dietrich, *Contingency Planning.*
34. Norcross, *Hurricane Almanac 2006*, 10–11.

## NOTES TO CHAPTER 7

1. The river is called the Souris in Canada and by the U.S. federal government, and the Mouse by North Dakotans.
2. Schramm, "The Day the Sirens Sounded Minot Dispatch."
3. Federal Emergency Management Agency (FEMA), "2011 Souris Valley Flood."
4. City of Minot, North Dakota, "Community Development Block Grant Disaster Recovery Action Plan."
5. Graney, "Minot's Undeniable Watermark Remains."
6. Reuters, "Minot Braces for Floods."

7. FEMA, "2011 Souris Valley Flood."
8. City of Minot, North Dakota, "2011 Mouse River Flood Response and Recovery."
9. City of Minot, North Dakota, "City of Minot Unmet Needs Assessment."
10. U.S. Census Bureau, Minot (city), North Dakota, QuickFacts, *http://quickfacts. census.gov/qfd/states/38/3853380.html.*
11. http://mds.mennonite.net/projects/minotnd/.
12. Jenkinson, "The 2012 Election in ND."
13. Hoeven, "Lessons from North Dakota's Oilfields."
14. Jordan Weissmann, "How Oil Made Working-Class North Dakota a Whole Lot Richer."
15. United States Geological Survey, "USGS Releases Updated Bakken . . ."
16. Jerry Chavez, the president of the Minot Area Development Corp., quoted in Wetzel and Foley, "North Dakota City Struggles with Flood Recovery."
17. Francis, "Who Is Looking Out for Minot?"
18. Centers for Disease Control and Prevention, "Suicide Among Adults . . ."
19. Schramm, "Crime, Housing, Flood Control."
20. McChesney, "Oil Boom Puts Strain . . ."
21. City of Minot, North Dakota, "Annual Budget, 2013."
22. Wetzel and Foley, "North Dakota City Struggles . . ."
23. "Hoeven: Minot Making Progress With CDBG Recovery Funds."
24. FEMA, "FEMA Provides $2 Million for Minot Floodwall Project."
25. McKone, "FEMA Announces Funding . . ."
26. Fundingsland, "Recovery Distant."
27. Falcon, "A Long Way to Go."
28. "Oak Park."
29. Johnson, "Fundraiser Dance Party . . ."
30. Johnson, "Staybridge Suites . . ."
31. "The week That Was."
32. Schramm, "City Seeks Volunteer Planners . . ."
33. Schramm, "Recovery Committee Reviews ESF-14 Progress."
34. Schramm, "Slow but Steady."
35. "City Faces Difficult Choice."
36. City of Minot, North Dakota, "2011 Mouse River Flood Response and Recovery."
37. FEMA, "Hope Village."
38. Ogden, "Minot Amtrak Depot . . ."
39. Davidian, "Levees Hold as Souris River Crests."
40. Chabun, "Border Flood Talks Begin."
41. Tran, "Minot Changes Flood Model."
42. Interview with local official, Minot, ND, July 24, 2012.
43. U.S. Census Bureau, 2011 MSA Business Patterns (NAICS), Minot, ND Metropolitan Statistical Area.
44. Ondracek and Witwer, "Minot, North Dakota Business Recovery Survey."
45. Ibid.
46. City of Minot, North Dakota, "2011 Mouse River Flood Response and Recovery."
47. U.S. Census Bureau, 2011 MSA Business Patterns (NAICS), Minot, ND Metropolitan Statistical Area.
48. Ibid.
49. Ibid.
50. Ibid.
51. Birks and Mills, *Grounded Theory,* 132.
52. Eisler, *False Expectations.*

53. Marquart, *The Horizontal World.*
54. Norris, *Dakota,* 113.
55. Norris, *Dakota,* 50.
56. Caldwell, "Are We Ready to Rumble?"
57. Johannisson and Olaison, "The Moment of Truth," 72–3.
58. Johnston, Becker, and Paton, "Multi-Agency Community Engagement . . ." 263.
59. Petrolia, Landry, and Coble, "Risk Preferences, Risk Perceptions, and Flood Insurance," 242.
60. "Delegation, Governor Announce More than $40 Million."
61. Brewton, Danes, Stafford, and Haynes, "Determinants of Rural and Urban Family Firm Resilience." See also Besser, "The Significance of Community to Business Social Responsibility," quoted within the cited text.
62. Southwick and Charney, *Resilience.*
63. Bligh and Wendelbo, "Hydraulic Fracturing," 7.
64. Hall, "Hydraulic Fracturing—a Primer," S2.
65. Abeyratne, "The Deepwater Horizon Disaster."
66. Gibson, "Bombing North Dakota: Living Amid the Bakken Oil Boom."
67. Karaim, "Welcome to the Boomtown."
68. Schmidt, "Blind Rush?" See also Jackson et al., "Increased Stray Gas Abundance . . ."
69. Fershee, "North Dakota Expertise."
70. Cook et al., "Quantifying the consensus . . ."
71. Hill and Olson, "Possible Future Trade-Offs," 311.
72. Ibid, 313.
73. Ibid.

## NOTES TO CHAPTER 8

1. Seawright and Gerring, "Case Selection Techniques in Case Study Research."
2. Ibid, 301.
3. Ibid, 299.
4. Ibid, 303.
5. Noy, "The Macroeconomic Consequences of Disasters," 229.
6. Ibid, 227.

# Bibliography

"Beacon Council Conducts Local Business Needs and Damage Assessment Survey Resulting from Hurricane Wilma." *Business Wire,* October 26, 2005.

"Chiefs, Dolphins Get Jump on Wilma." *Buffalo News,* October 21, 2005.

"Citibank Reopens Bank Branches in Miami-Dade County; Several Financial Centers Fully Operational." *Business Wire,* October 26, 2005.

"City Faces Difficult Choice." *Minot Daily News,* December 12, 2012.

"Delegation, Governor Announce More than $40 Million in Additional CDBG-DR Disaster Aid for Minot, Ward County." *Targeted News Service,* March 27, 2013.

"Hoeven: Minot Making Progress With CDBG Recovery Funds." *America's Intelligence Wire,* May 3, 2013.

"Mother's Day Second-Line Shooting on Frenchmen Street Injures at Least 19 People." *The Times-Picayune,* May 12, 2013.

"Oak Park: The Check Is in Minot." *Minot Daily News,* October 25, 2011.

"The Week That Was." *Minot Daily News,* May 13, 2012.

"Wilma's Impact on the Area." *Palm Beach Post,* October 24, 2005.

Abeyratne, Ruwantissa. "The Deepwater Horizon Disaster—Some Liability Issues." *Tulane Maritime Law Journal* 35, no. 1 (2010), 125.

Adams, Guy B. and Bayard L. Catron. "Communitarianism, Vickers, and Revisioning American Public Administration." *The American Behavioral Scientist* 38, no. 1 (1994), 44–63.

Adger, W. Neil. "Social and Ecological Resilience: Are They Related?" *Progress in Human Geography* 24, no. 3 (2000), 347–364.

Alesch, Daniel J., Lucy A. Arendt, and James N. Holly. *Managing for Long-Term Community Recovery in the Aftermath of Disaster.* Fairfax, VA: Public Entity Risk Institute, 2009.

Altay, Nezih and Andres Ramirez. "Impact of Disasters on Firms in Different Sectors: Implications for Supply Chains." *Journal of Supply Chain Management* 46, no. 4 (2010), 59–80.

Anderson, Benedict R. O'G. *Imagined Communities: Reflections on the Origin and Spread of Nationalism.* London: Verso, 1991.

Anderson, Curt. "Wilma's Wrath: Deadly Hurricane Buffets Florida." *Deseret News,* October 25, 2005.

Arakida, Masaru. "Measuring Vulnerability: The ADRC Perspective for the Theoretical Basis and Principles of Indicator Development." *Measuring Vulnerability to Natural Hazards—towards Disaster Resilient Societies.* Tokyo: United Nations University (2006), 290–299.

Archer, Diane and Somsook Boonyabancha. "Seeing a Disaster as an Opportunity—Harnessing the Energy of Disaster Survivors for Change." *Environment and Urbanization* 23, no. 2 (2011), 351–364.

Associated Press. "Gov. Bush Criticizes State's Storm Effort; He Says Florida, Not FEMA, Is to Blame." *The Washington Post,* October 27, 2005.

Atkinson, Christopher L. "An Evaluation of the Impact of Local Government Institutions on Business Resilience in Disaster." PhD diss., Florida Atlantic University, 2011. http://digitool.fcla.edu/dtl_publish/33/3174502.html.

———. "Small Business Development: A Comparison of Programs in American Cities and Counties." *The Journal of Contemporary Issues in Business and Government,* 17, no. 2 (2011), 63–84.

Atkinson, Christopher L. and Alka K. Sapat. "After Katrina: Comparisons of Post-Disaster Public Procurement Approaches and Outcomes in the New Orleans Area." *Journal of Public Procurement,* 12, no. 3 (2012), 356–385.

Atkinson, Rowland and Gary Bridge. *Gentrification in a Global Context: The New Urban Colonialism.* London: Routledge, 2005.

Bankoff, Greg. "The Historical Geography of Disaster: 'Vulnerability' and 'Local Knowledge' in Western Discourse." In *Mapping Vulnerability: Disasters, Development and People,* edited by Greg Bankoff, Georg Frerks, and Dorothea Hilhorst. London: Earthscan, 2004.

Barnshaw, John and Joseph Trainor. "Race, Class, and Capital amidst the Hurricane Katrina Diaspora." In *The Sociology of Katrina: Perspectives on a Modern Catastrophe,* edited by David L. Brunsma, David Overfelt, and J. Steven Picou. Lanham, MD: Rowman & Littlefield, 2007.

Bea, Keith. *Federal Disaster Policies After Terrorists Strike: Issues and Options.* Hauppauge, NY: Novinka Books, 2003.

Berke, Philip R., Jack Kartez, and Dennis Wenger. "Recovery after Disaster: Achieving Sustainable Development, Mitigation, and Equity." *Disasters* 17, no. 2 (1993), 93–109.

Besser, Terry L. "The Significance of Community to Business Social Responsibility." *Rural Sociology* 63, no. 3 (1998), 412–31.

Birks, Melanie and Jane Mills. *Grounded Theory: A Practical Guide.* London: Sage, 2011.

Blackburn, Robert A. and Michael T. Schaper. *Government SMEs and Entrepreneurship Development: Policy Practice and Challenges.* Farnham: Gower, 2012.

Blake, Eric S., Christopher W. Landsea, and Ethan J. Gibney. "The Deadliest, Costliest, And Most Intense United States Tropical Cyclones From 1851 To 2010 (And Other Frequently Requested Hurricane Facts)." (2011, August). http://www.nhc.noaa.gov/pdf/nws-nhc-6.pdf.

Blanco, Magdalena P. "Preliminary Assessment of Statutory Compliance of 4A and 4B Economic Development Corporations in Texas with the Development Corporation Act of 1979" (2009). Applied Research Projects, Texas State University-San Marcos. https://digital.library.txstate.edu/handle/10877/3795.

Bligh, Shawna and Chris Wendelbo. "Hydraulic Fracturing: Drilling into the Issue." *Natural Resources & Environment* 27, no. 3 (2013), 7.

Blomley, Nicholas K. *Unsettling the City: Urban Land and the Politics of Property.* New York: Routledge, 2004.

Boodhoo, Niala. "Forget Wilma and Focus on Area's Strengths, Experts Say." *The Miami Herald—Business,* November 28, 2005. http://www.wraggcasas.com/news/forget_wilma.html.

Boyer, M. Christine. "Cities for Sale: Merchandising History at South Street Seaport." In *Variations on a Theme Park: The New American City and the End of Public Space,* edited by Michael Sorkin. New York: Hill and Wang, 1992.

Brand, Fridolin Simon and Kurt Jax. "Focusing the Meaning(s) of Resilience: Resilience as a Descriptive Concept and a Boundary Object." *Ecology and Society* 12, no. 1 (2007), 23. http://www.ecologyandsociety.org/vol12/iss1/art23/.

Brandeis, Louis Dembitz. *Business—A Profession.* Boston: Small, Maynard, & Company, 1914.

Brewton, Katherine E., Sharon M. Danes, Kathryn Stafford, and George W. Haynes. "Determinants of Rural and Urban Family Firm Resilience." *Journal of Family Business Strategy,* 1 (2010), 155–166.

Briseno, Vanessa. "A Chronicle of My First Trip to the Remarkable and Resilient City of New Orleans." *Global Green Blog,* April 23, 2013. http://globalgreen. org/blogs/global/wp-trackback.php?p = 6728.

Brown, Stephen P. A., Kathy J. Hayes, and Lori L. Taylor. "State and Local Policy, Factor Markets, and Regional Growth." *Review of Regional Studies* 33, no. 1 (2003), 40–60.

Browning, Robert. *Men and Women and Sordello,* Boston: Houghton, Mifflin, and Company, 1886. Google e-book.

Bureau of Governmental Research. "Contracting with Confidence: Professional Services Contracting Reform in New Orleans." (2010). http://www.bgr.org/files/ reports/Contracting_w_Confidence.pdf.

———. "Private Services in the Public Interest: Reforming Jefferson Parish's Unusual Approach to Service Contracting." (2012). http://www.bgr.org/files/reports/BGR_ Contracting-Jeff.pdf.

Buttress, Steve and Don Macke. "Energizing Entrepreneurs." *Economic Development Journal* 7, no. 4 (2008), 20–25.

Caldwell, Dave. "Are we ready to rumble? History, city officials provide good news." *Minot Daily News,* January 25, 2009.

Callegaro, Mario and Charles Disogra. "Computing Response Metrics for Online Panels." *The Public Opinion Quarterly* 72, no. 5 (2008), 1008–32.

Centers for Disease Control and Prevention. "Suicide Among Adults Aged 35–64 Years—United States, 1999–2010" *Morbidity and Mortality Weekly Report* 62, no. 17 (2013, May 3), 321–325.

Chabun, Will. "Border Flood Talks Begin." *The Leader-Post,* March 23, 2013.

Childs, Donna R. and Stefan Dietrich. *Contingency Planning and Disaster Recovery: A Small Business Guide.* Hoboken, NJ: John Wiley, 2002.

City of Minot, North Dakota. "2011 Mouse River Flood Response and Recovery (Community Press Kit)." June 22, 2012. http://www.minotrecoveryinfo. com/uploads%5Cresources%5C65%5Ccofm_community-press-kit_one-year-later_061912.pdf

———. "Annual Budget, 2013." http://www.minotnd.org/pdf/finance/budget/13-final. pdf.

———. "City of Minot Unmet Needs Assessment: Helping the City of Minot Recover from the Mouse River Flood of 2011, October 16, 2012." http://www. minotrecoveryinfo.com/uploads%5Cresources%5C80%5Ccofm_unmet-needs-assessment_110712.pdf.

———. "Community Development Block Grant Disaster Recovery Action Plan, July 12, 2012." http://www.minotrecoveryinfo.com/uploads/resources/70/cofm_ cdbgdr-final-action-plan_7.16.12.pdf.

Clandinin, D. Jean, Janice Huber, Marilyn Huber, M. Shaun Murphy, Anne Murray Orr, Marni Pearce, and Pam Steeves. *Composing Diverse Identities: Narrative Inquiries into the Interwoven Lives of Children and Teachers.* New York: Routledge, 2006.

Clawson, Dan, Alan Neustadtl, and Mark Weller. *Dollars and Votes: How Business Campaign Contributions Subvert Democracy.* Philadelphia, PA: Temple University Press, 1998.

Clingermayer, James C. and Richard Feiock. *Institutional Constraints and Policy Choice: An Exploration of Local Governance.* Albany, NY: SUNY, 2001.

Colten, Craig E., Robert W. Kates, and Shirley B. Laska. "Community Resilience: Lessons From New Orleans and Hurricane Katrina." CARRI Research Report 3, Oak Ridge National Laboratory, 2008.

Constantino, Nicola, Mariagrazia Dotoli, Marco Falagario, and Maria Pia Fanti. "Using Fuzzy Decision Making for Supplier Selection in Public Procurement." *Journal of Public Procurement,* 11, no. 3 (2011), 403–427.

Cook, John, Dana Nuccitelli, Sarah A Green, Mark Richardson, Bärbel Winkler, Rob Painting, Robert Way, Peter Jacobs, and Andrew Skuce. "Quantifying the Consensus on Anthropogenic Global Warming in the Scientific Literature." *Environmental Research Letters* 8, no. 2 (2013). http://dx.doi.org/10.1088/1748-9326/8/2/024024.

Cortés, Alfons and Salvatore Rizzello. "Hayek's Theory of Knowledge and Behavioural Finance." In *Cognition and Economics (Advances in Austrian Economics, Volume 9),* edited by Elisabeth Krecké, Carine Krecké, and Roger G. Koppl. Amsterdam: Emerald, 2006.

Creswell, John W. *Research Design: Qualitative, Quantitative, and Mixed Methods Approaches, 3rd ed.* Thousand Oaks, CA: Sage, 2009.

Crutcher, Michael E., Jr. *Tremé: Race and Place in a New Orleans Neighborhood.* Athens: University of Georgia Press, 2010.

Cutter, Susan L. *Living with Risk.* London: Edward Arnold, 1993.

Dallimore, Suzanne M. "The Ten Commandments of Ethical Government Purchasing." *Government Procurement* 14, no. 2 (2006), 28–29.

Dash, Nicole, Walter Gillis Peacock and Betty Hearn Morrow. "And the Poor Get Poorer: A Neglected Black Community." In *Hurricane Andrew: Ethnicity, Gender and the Sociology of Disasters,* edited by Walter Gillis Peacock, Betty Hearn Morrow and Hugh Gladwin. New York: Routledge, 1997.

Davidian, Geoff. "Levees Hold as Souris River Crests; Officials Credit Efforts in Canada to Hold Back Flood Waters in Reservoirs Along Waterway." *The Vancouver Sun,* June 27, 2011.

Davis, Mike. "Gentrifying Disaster." *Mother Jones,* October 25, 2005. http://www.motherjones.com/politics/2005/10/gentrifying-disaster.

Davis, Tom (Chairman). *A Failure of Initiative: The Final Report of the Select Bipartisan Committee to Investigate the Preparation for and Response to Hurricane Katrina.* Select Bipartisan Committee to Investigate the Preparation for and Response to Hurricane Katrina (2006). http://katrina.house.gov/.

De Mel, Suresh, David McKenzie, and Christopher Woodruff. "Enterprise Recovery Following Natural Disasters." *The Economic Journal* 122, no. 559 (2011), 64–91.

Deephouse, David L. and Mark Suchman. "Legitimacy in Organizational Institutionalism." In *The SAGE Handbook of Organizational Institutionalism,* edited by Royston Greenwood, Christine Oliver, Roy Suddaby, and Kerstin Sahlin. London: Sage, 2008.

Dekker, Sidney, Richard Cook, Leila Johannesen, and Nadine Sarter. *Behind Human Error, 2nd ed.* Farnham: Ashgate, 2010.

Denhardt, Robert B. and Janet Vinzant Denhardt. "The New Public Service: Serving Rather Than Steering." *Public Administration Review* 60, no. 6 (2000), 549–59.

Denzin, Norman K. and Yvonna S. Lincoln. *Handbook of Qualitative Research.* Thousand Oaks, CA: Sage, 1994.

DiMaggio, Paul J. and Walter W. Powell. "The Iron Cage Revisited: Institutional Isomorphism and Collective Rationality in Organizational Fields." *American Sociological Review,* 48, no. 2 (1983), 147–160.

Dominguez, Jorge I. *Economic Strategies and Policies in Latin America.* New York: Garland, 1994.

Donahue, Amy K. and Sean O'Keefe. "Universal Lessons from Unique Events: Perspectives from Columbia And Katrina." *Public Administration Review* 67, no. 6 (2007), 77–81.

Douthat, Bill and Jane Musgrave. "Officials from State to Local Levels Join Forces." *Palm Beach Post,* October 20, 2005.

Dovers, Stephen and John Handmer. *The Handbook of Disaster and Emergency Policies and Institutions.* London: Earthscan, 2007.

Edwards, Frances L. "Obstacles to Disaster Recovery." *The Public Manager,* Winter, 2008–2009.

Eisler, Dave. *False Expectations: Politics and the Pursuit of the Saskatchewan Myth.* Regina, SK: Canadian Plains Research Center, 2006.

Elhauge, Einger R. "Does Interest Group Theory Justify More Intrusive Judicial Review." *The Yale Law Journal* 101, no. 31 (1991), 32.

Elkin, Stephen L. "State and Market in City Politics: Or, The 'Real Dallas.'" In *The Politics of Urban Development,* edited by Clarence N. Stone and Heywood T. Sanders. Lawrence: University Press of Kansas, 1987.

Elran, Meir. "Measuring Societal Resilience." *The Proceedings of the First International Symposium on Societal Resilience, November 30–December 2, 2010, Fairfax, Virginia.* http://homelandsecurity.org/docs/Social_Resilience_BOOK.pdf.

Enarson, Elaine and Betty Hearn Morrow. "Women Will Rebuild Miami: A Case Study of Feminist Response to Disaster." In *The Gendered Terrain of Disaster: Through Women's Eyes,* edited by Elaine Enarson and Betty Hearn Morrow. Westport, CT: Praeger, 1998.

Esnard, Ann-Margaret, Alka Sapat, and Diana Mitsova. "An Index of Relative Displacement Risk to Hurricanes." *Natural Hazards* 59, no. 2 (2011), 833–59.

Ewing, Bradley T., Jamie Brown Kruse, and Dan Sutter. "Hurricanes and Economic Research: An Introduction to the Hurricane Katrina Symposium." *Southern Economic Journal* 74, no. 2 (2007), 315–25.

Falcon, James C. "A Long Way to Go: Oak Park Recovery, Restoration Expected to Be a Lengthy Process." *Minot Daily News,* September 13, 2011.

Farber, Daniel A. and Jim Chen. *Disasters And the Law: Katrina and Beyond.* New York: Aspen Publishers, 2006.

Faust, Kelly L. and Susan M. Carlson. "Devastation in the Aftermath of Hurricane Katrina as a State Crime: Social Audience Reactions." *Crime, Law and Social Change* 55, no. 1 (2011), 33–51.

Federal Emergency Management Agency (FEMA). "2011 Souris Valley Flood: By the Numbers." 2012. http://www.fema.gov/2011-souris-valley-flood-numbers.

———. "FEMA Provides $2 Million For Minot Floodwall Project." March 19, 2013. http://www.fema.gov/news-release/2013/03/19/fema-provides-2-million-minot-floodwall-project.

———. "Hope Village: Helping Volunteers Help the Souris Valley." June 11, 2012. http://www.fema.gov/news-release/2012/06/11/hope-village-helping-volunteers-help-souris-valley.

———. "By the Numbers: First 100 days—FEMA Recovery Update for Hurricane Katrina." December 7, 2005. http://reliefweb.int/report/united-states-america/numbers-first-100-days-fema-recovery-update-hurricane-katrina.

Fershee, Joshua P. "North Dakota Expertise: A Chance to Lead in Economically and Environmentally Sustainable Hydraulic Fracturing." *North Dakota Law Review* 87, no. 4 (2011), 485.

Fincher, Ruth-Marie and Jane Margaret Jacobs. *Cities of Difference.* New York: Guilford, 1998.

Fink, Steven. *Crisis Management: Planning for the Inevitable.* New York: American Management Association, 1986.

Fleischman, Arnold and Gary P. Green. "Organizing Local Agencies to promote Economic Development." *The American Review of Public Administration* 21, no. 1 (1991), 1–15.

Florida Power & Light (FPL). "FPL's First Full Day of Restoration Gets Underway; Assessment Will Identify Extent of Damage." October 25, 2005. http://www.fpl.com/news/2005/contents/05170.shtml.

———. "Remaining 14 Percent without Power Post-Wilma is Focus of Restoration Team from 33 States and Canada." *Business Wire,* November 4, 2005.

Folke, Carl. "Resilience: The Emergence of a Perspective for Social–Ecological Systems Analyses." *Global Environmental Change* 16, no. 3 (2006), 253–267.

Fothergill, Alice. "The Neglect of Gender in Disaster Work: An Overview Of The Literature." In *The Gendered Terrain of Disaster: Through Women's Eyes,* edited by Elaine Enarson and Betty Hearn Morrow. Westport, CT: Praeger, 1998.

Fowler, Floyd J., Jr. *Survey Research Methods.* Newbury Park, CA: Sage, 1988.

Frailing, Kelly and Dee Wood Harper. "Crime and Hurricanes in New Orleans." In *The Sociology of Katrina: Perspectives on a Modern Catastrophe,* edited by David L. Brunsma, David Overfelt, and J. Steven Picou. Lanham, MD: Rowman & Littlefield, 2007.

Francis, Rhonda. "Who Is Looking Out for Minot?" *Minot Daily News,* April 28, 2013.

Frogameni, Bill. "Florida on South Florida: Housing Bust 'Created Opportunities.'" *South Florida Business Journal,* December 10, 2010.

Fundingsland, Kim. "Recovery Distant: Minot parks, Zoo Flood Recovery Date Inestimable." *Minot Daily News,* January 27, 2012.

Gallopín, Gilberto C. "Linkages between Vulnerability, Resilience, and Adaptive Capacity." *Global Environmental Change* 16, no. 3 (2006), 293–303.

Gephart, Robert P. , Jr. "Crisis Sensemaking and the Public Inquiry." In *International Handbook of Organizational Crisis Management,* edited by Christine M. Pearson, Christophe Roux-Dufort and Judith A. Clair, 123–61. Thousand Oaks, CA: Sage, 2007.

Gerstein, Marc S. and Michael Ellsberg. *Flirting With Disaster: Why Accidents Are Rarely Accidental.* New York: Sterling, 2008.

Gibson, James William. "Bombing North Dakota: Living Amid the Bakken Oil Boom." *Earth Island Journal* 27, no. 4 (2013), 30.

Gilbert, Claude. "Studying Disaster: Changes in the Main Conceptual Tools." In *What is a Disaster? A Dozen Perspectives on the Question,* edited by E. L. Quarantelli. London: Routledge, 1998.

Graney, Emma. "Minot's Undeniable Watermark Remains." *The Leader-Post,* March 15, 2013.

Grazulis, Thomas P. *Significant Tornadoes, 1680–1991.* St. Johnsbury, VT: Environmental Films, 1993.

Greenwood, Royston, Christine Oliver, Roy Suddaby, and Kerstin Sahlin. "Introduction." *The SAGE Handbook of Organizational Institutionalism.* London: Sage, 2008.

Gupta, Rani and Eliot Kleinberg. "Hardest-Hit Cities Get Handle on Cost of Wilma's Damage." *Palm Beach Post,* October 29, 2005.

Hall, Keith B. "Hydraulic Fracturing—a Primer." *The Enterprise* 41, no. 11 (Oct 10, 2011), S2.

Halter, Maria Virginia, Maria Cecilia Coutinho de Arruda, and Ralph Bruno Halter. "Transparency to Reduce Corruption?" *Journal of Business Ethics* 84, no. 3 (2009), 373–85.

Haltiwanger, John and C. J. Krizan. "Small Business and Job Creation in the United States: The Role of New and Young Businesses." In *Are Small Firms Important?: Their Role and Impact,* edited by Zoltán J. Ács. Boston: Kluwer, 1999.

Haltiwanger, John, Ron S. Jarmin, and Javier Miranda. "Who Creates Jobs? Small vs. Large vs. Young." 2011. http://econweb.umd.edu/~haltiwan/size_age_paper_R&R_Aug_16_2011.pdf.

Handal, Jane. "Rebuilding City Identity through History: The Case of Bethlehem-Palestine." In *Designing Sustainable Cities in the Developing World,* edited by Roger Zetter and Georgia Butina Watson. Aldershot: Ashgate, 2006.

Harris, Robert. *Political Corruption: In and Beyond the Nation State.* London: Routledge, 2003.

Hawkins, Christopher V. and Richard Feiock, "Joint Ventures, Economic Development Policy, and the Role of Local Governing Institutions." *The American Review of Public Administration* 41, no. 3 (2011), 329–47.

Hayek, Friedrich A. *The Constitution Of Liberty.* Chicago: University of Chicago, 1960.

Heisler, Paul K. "A Theoretical Comparison of Certified Piano Teachers' Claim to Professional Status with the Sociological Definition of Profession." *International Review of the Aesthetics and Sociology of Music* 26, no. 2 (1995), 239–49.

Hewitt, Kenneth. "Excluded Perspectives in the Social Construction of Disaster." *International Journal of Mass Emergencies and Disasters* 13, no. 3 (1995), 317–39.

Hill, Michael J. and Rhonda Olson. "Possible Future Trade-offs between Agriculture, Energy Production, and Biodiversity Conservation in North Dakota." *Regional Environmental Change* 13, no. 2 (2013), 311–28.

Hobbes, Thomas. *Leviathan, or, the Matter, Form, and Power of a Commonwealth, Ecclesiastical and Civil, 2nd ed.* London: George Routledge, 1886.

Hoeven, John. "Lessons from North Dakota's Oilfields." *The American Spectator,* February, 2013. http://spectator.org/archives/2013/02/14/lessons-from-north-dakotas-oil.

Holling, C. S. "Resilience and Stability of Ecological Systems." *Annual Review of Ecology and Systematics,* 4 (1973), 1–23.

Horlick-Jones, Tom. "Modern Disasters as Outrage and Betrayal." *International Journal of Mass Emergencies and Disasters* 13, no. 3 (1995), 305–15.

Huriash, Lisa J. "Wilma Wound Still Hurts." *South Florida Sun—Sentinel,* December 9, 2012.

Ingram, Paul and Karen Clay. "The Choice-within-Constraints New Institutionalism and Implications for Sociology." *Annual Review of Sociology* 26, no. 1 (2000), 525–546.

Irvine, Wilson and Alistair R. Anderson. "Small Tourist Firms in Rural Areas: Agility, Vulnerability and Survival in the Face of Crisis." *International Journal of Entrepreneurial Behaviour & Research* 10, no. 4 (2004), 229–46.

Jackson, Michael. "Responsibility versus Accountability in the Friedrich-Finer Debate." *Journal of Management History* 15, no. 1 (2009), 66–77.

Jackson, Robert B., Avner Vengosh, Thomas H. Darrah, Nathaniel R. Warner, Adrian Down, Robert J. Poreda, Stephen G. Osborn, Kaiguang Zhao, and Jonathan D. Karr. "Increased Stray Gas Abundance in a Subset of Drinking Water Wells near Marcellus shale Gas Extraction." *Proceedings of the National Academy of Sciences of the United States of America* (2013), doi: 10.1073/pnas.1221635110.

Jacobs, Jane M. "Staging Difference: Aestheticization and the Politics of Difference in Contemporary Cities." In *Cities of Difference,* edited by Ruth-Marie Fincher and Jane Margaret Jacobs. New York: Guilford, 1998.

Janis, Irving L. *Groupthink: Psychological Studies of Policy Decisions and Fiascoes.* 2nd ed. Boston: Houghton Mifflin, 1982.

Janssen, Marco A. and Elinor Ostrom. "Resilience, Vulnerability, and Adaptation: A Cross-Cutting Theme of the International Human Dimensions Programme on Global Environmental Change." *Global Environmental Change* 16, no. 3 (2006), 237–239.

Jenkinson, Clay. "The 2012 Election in ND: A Resounding Endorsement of Oil Development." *The Bismarck Tribune,* November 18, 2012.

Jennings, Marianne M. *Business: Its Legal, Ethical And Global Environment, 6th ed.* Mason, OH: Thomson/South Western, 2003.

Johannisson, Bengt and Lena Olaison. "The Moment of Truth—Reconstructing Entrepreneurship and Social Capital in the Eye of the Storm." *Review of Social Economy* 65, no. 1 (2007), 55–78.

Johnson, Andrea. "Fundraiser Dance Party to Benefit the Roosevelt Park Zoo." *Minot Daily News,* November 28, 2012.

Johnson, Andrea. "Staybridge Suites Volunteers Help Clean Up Roosevelt Zoo." *Minot Daily News,* August 12, 2012.

Johnston, David, Julia Becker, and Douglas Paton. "Multi-Agency Community Engagement during Disaster Recovery: Lessons from Two New Zealand Earthquake Events." *Disaster Prevention and Management* 21, no. 2 (2012), 252–268.

Karaim, Reed. "Welcome to the Boomtown." *Architect* 101, no. 7 (2012), 120–136.

Kates, Robert W., Craig E. Colten, Shirley B. Laska, and Stephen P. Leatherman. "Reconstruction of New Orleans after Hurricane Katrina: A Research Perspective." *Proceedings of the National Academy of Sciences of the United States of America* 103, no. 40 (2006), 14653–14660.

Keim, Barry D. and Robert A. Muller. *Hurricanes of the Gulf of Mexico.* Baton Rouge, LA: Louisiana State University Press, 2009.

Kettl, Donald F. and James W. Fesler. *The Politics of the Administrative Process, 3rd ed.* Washington, DC: CQ Press, 2005.

Keyes, W. Noel. *Government Contracts Under the Federal Acquisition Regulation, 3rd ed.* St. Paul, MN: Thomson/West, 2003.

Knabb, Richard D., Jamie R. Rhome, and Daniel P. Brown. "Tropical Cyclone Report, Hurricane Katrina, 23–30 August 2005" (September 14, 2011 update).

Kolb, Robert W. *Encyclopedia of Business Ethics And Society, Volume 2.* Thousand Oaks: Sage, 2008.

Kovács, Attila. *Enhancing Procurement Practices: Comprehensive Approach to Acquiring Complex Facilities and Projects.* Norwell, MA: Kluwer, 2004.

Krishna, Sankaran. *Globalization and Postcolonialism: Hegemony and Resistance in the Twenty-First Century.* Lanham, MD: Rowman & Littlefield, 2009.

Kunreuther, Howard. "Risk and Reaction." *Harvard International Review* 28, no. 3 (2006), 38–42.

Landström, Hans. *Pioneers in Entrepreneurship and Small Business Research.* New York: Springer, 2005.

Leicht, Kevin T. and Mary L. Fennell. "Institutionalism and the Professions." In *The SAGE Handbook of Organizational Institutionalism,* edited by Royston Greenwood, Christine Oliver, Roy Suddaby, and Kerstin Sahlin. London: Sage, 2008.

Leong, Karen J., Christopher A. Airriess, Wei Li, Angela Chia-Chen Chen, and Verna M. Keith. "Resilient History and the Rebuilding of a Community: The Vietnamese American Community in New Orleans East." *The Journal of American History* 94, no. 3 (2007), 770–9.

LeSage, James, R. Kelley Pace, Richard Campanella, Nina Lam, and Xingjian Liu. "Do What the Neighbours Do: Reopening Businesses after Hurricane Katrina." *Significance* 8, no. 4 (2011), 160–3.

Levitt, Jeremy I. and Matthew C. Whitaker. *Hurricane Katrina: America's Unnatural Disaster.* Lincoln: University of Nebraska, 2009.

Lewis, Peirce F. *New Orleans: The Making of an Urban Landscape, second edition.* Charlottesville, VA: University of Virginia Press, 2003.

Light, Ivan and Steven J. Gold. *Ethnic Economies.* San Diego: Academic Press, 2000.

Lipman, Pauline. "The Cultural Politics of Mixed-Income Schools and Housing: A Racialized Discourse of Displacement, Exclusion, and Control." *Anthropology & Education Quarterly* 40, no. 3 (2009), 215–36.

Liu, Baodong, James M. Vanderleeuw. *Race Rules: Electoral Politics in New Orleans, 1965–2006.* Lanham, MD: Rowman & Littlefield, 2007.

Locke, John. *Two Treatises of Government.* London: C. and J. Rivington, 1824. Google e-book.

Long, Jerrold A. "Overcoming Neoliberal Hegemony in Community Development: Law, Planning, and Selected Lamarckism." *The Urban Lawyer* 44, no. 2 (2012), 345+. LegalTrac.

Lynn, Dahlia Bradshaw. "Succession Management Strategies in Public Sector Organizations: Building Leadership Capital." *Review of Public Personnel Administration* 21, no. 2 (2001), 114–32.

Macfarlan, W. Gregor. "From Philosophy to Practice: A Case of Acquisition Management Reform." *The Public Manager: The New Bureaucrat* 27, no. 1 (1998), 7.

Mantzavinos, C. *Individuals, Institutions, and Markets*. Cambridge: Cambridge University Press, 2001.

March, James G. and Johan P. Olsen. "Elaborating the New Institutionalism." In *The Oxford Handbook of Political Science*, edited by Robert E. Goodin. Oxford: Oxford University Press, 2009.

———. "The New Institutionalism: Organizational Factors In Political Life." *The American Political Science Review* 78, no. 3 (1984), 734–49.

Marcus, Noreen and Sally Kestin. "Deadliest Hurricane Category: Cleanup Accidents Kill More than Winds from Storm." *South Florida Sun—Sentinel*, Nov 8, 2005.

Marquart, Debra. *The Horizontal World: Growing Up Wild in the Middle of Nowhere*. New York: Counterpoint, 2006.

Martin, Michael G. *Federal Statements of Work: A Practical Guide*. Vienna, VA: Management Concepts, 2008.

Martin, Naomi. "After 'Tragic' Mother's Day Mass Shooting, Mayor Mitch Landrieu Urges Community to Rally against Violence." *The Times-Picayune*, May 12, 2013.

Martin, Naomi. "New Orleans 'Shell Shocked' Documentary Examines Grim Reality of Murder through the Eyes of Children." *The Times-Picayune*, May 2, 2013.

Mayk, Lauren and Cathy Zollo. "Wilma's Destruction Baffles FPL Officials; the Hurricane Left 3.2 Million in the Dark and Knocked Down 10,000 Poles, More than any of the State's Recent Storms." *Sarasota Herald Tribune*, October 28, 2005.

McChesney, John. "Oil Boom Puts Strain On North Dakota Towns." *NPR*, December 2, 2011. http://www.npr.org/2011/12/02/142695152/oil-boom-puts-strain-on-north-dakota-towns.

McKone, Tommy. "FEMA Announces Funding Approval for Phase I of Minot Water Treatment Plant Protection." Press release, March 19, 2013. http://cramer.house.gov/media-center/press-releases/fema-announces-funding-approval-for-phase-i-of-minot-water-treatment.

McNabb, David E. *Research Methods in Public Administration and Nonprofit Management: Quantitative and Qualitative Approaches, 2nd ed.* Armonk, NY: M.E. Sharpe, 2008.

Meyer, John W and Brian Rowan. "Institutionalized Organizations: Formation Structure as Myth and Ceremony." *American Journal of Sociology* 83 (1977), 440–463.

———. "The Structure of Educational Organizations." (1983), quoted in "Introduction," *The SAGE Handbook of Organizational Institutionalism*, edited by Royston Greenwood, Christine Oliver, Roy Suddaby, and Kerstin Sahlin. London: Sage, 2008.

Miller, Dale T. "The Norm of Self-Interest." *American Psychologist* 54, no. 12 (1999), 1053–1060.

Miller, David L. *Principles of Social Justice*. Cambridge, MA: Harvard University Press, 1999.

Miller, DeMond Shondell and Jason David Rivera. *Community Disaster Recovery and Resiliency: Exploring Global Opportunities and Challenges*. Boca Raton, FL: CRC Press, 2011.

Miller, Hugh Theodore and Charles J. Fox. *Postmodern Public Administration, rev. ed.* Armonk, NY: M.E. Sharpe, 2007.

Mitchell, James K. *Crucibles of Hazard: Mega-Cities and Disasters in Transition.* New York: United Nations University, 1999.

Mitroff, Ian I. *Why Some Companies Emerge Stronger and Better from a Crisis: 7 Essential Lessons for Surviving Disaster.* New York: American Management Association, 2005.

Montesquieu, Charles-Louis de Secondat, Baron de La Brède et de and Melvin Richter. *The Political Theory of Montesquieu.* Cambridge: Cambridge University Press, 1977.

Mooney, Chris. *Storm World: Hurricanes, Politics, and the Battle over Global Warming.* Orlando, FL: Harcourt, 2007.

Munich RE. "Natural Catastrophe Statistics for 2012 Dominated by Weather Extremes in the USA." 2013. http://www.munichre.com/en/media_relations/press_releases/2013/2013_01_03_press_release.aspx.

———. "Review of Natural Catastrophes in 2011: Earthquakes Result in Record Loss Year." 2012. http://www.munichre.com/en/media_relations/press_releases/2012/2012_01_04_press_release.aspx.

Musgrave, Richard A. "A Multiple Theory of Budget Determination." *FinanzArchiv/ Public Finance Analysis* 17, no. 3 (1956), 333–43.

Nafziger, E. Wayne. *Economic Development, 4th ed.* Cambridge: Cambridge University Press, 2006.

National Hurricane Center. "2005 Atlantic Hurricane Season." http://www.nhc.noaa.gov/2005atlan.shtml.

National Research Council. *Building Community Disaster Resilience Through Private-Public Collaboration.* Washington, DC: National Academies Press, 2011.

Nee, Victor. "The New Institutionalisms in Economics and Sociology." In *The Handbook of Economic Sociology, 2nd ed.*, edited by Neil J. Smelser and Richard Swedberg. Princeton, NJ: Princeton University Press, 2005.

Niskanen Jr., William A. *Bureaucracy and Representative Government.* New Brunswick, NJ: Aldine Transaction, 2007.

Noon, Christopher R. "The Use of Racial Preferences in Public Procurement for Social Stability." *Public Contract Law Journal* 38, no. 3 (2009), 611.

Norcross, Bryan. *Hurricane Almanac 2006: The Essential Guide to Storms Past, Present, and Future.* New York: St. Martin's Press, 2006.

Norris, Kathleen. *Dakota: A Spiritual Geography.* New York: Ticknor & Fields, 1993.

North, Douglass Cecil. *Institutions, Institutional Change, and Economic Performance.* Cambridge: Cambridge University Press, 1990.

Noy, Ilan. "The macroeconomic consequences of disasters." *Journal of Development Economics* 88, no. 2 (2009), 221–31.

Oberle, Wayne H., Kevin R. Stowers, James P. Darby. "A Definition of Development." *Journal of the Community Development Society* 5 (1974), 61–71.

Ogden, Eloise. "Minot Amtrak Depot Back in Business." *Minot Daily News,* April 24, 2013.

O'Looney, John. *Outsourcing State and Local Government Services: Decision-Making Strategies and Management Methods.* Westport, CT: Quorum, 1998.

Olshansky, Robert B. and Laurie Johnson. *Clear as Mud: Planning for the Rebuilding of New Orleans.* Chicago: American Planning Association, 2010.

Ondracek, James and Keith Witwer. "Minot, North Dakota Business Recovery Survey II Final Report, prepared for Minot Area Development Corporation, 2012."

Orbach, Barak. *Regulation: Why and How the State Regulates.* New York: Foundation, 2013.

Ostrom, Elinor. "A Diagnostic Approach for Going Beyond Panaceas." *Proceedings of the National Academy of Sciences of the United States of America* 104, no. 39 (2007), 15181–7.

Partridge, Mark D., and Dan S. Rickman. "Do We Know Economic Development When We See It?" *Review of Regional Studies* 33, no. 1 (2003), 17–39.

Pasch, Richard J., Eric S. Blake, Hugh D. Cobb III, and David P. Roberts. "Tropical Cyclone Report, Hurricane Wilma, 15–25 October 2005." National Hurricane Center (2006).

Paton, Douglas and David Moore Johnston. *Disaster Resilience: An Integrated Approach*. Springfield, IL: Charles C Thomas, 2006.

Pavlichev, Alexei and G. David Garson. *Digital Government: Principles and Best Practices*. Hershey, PA: Idea Group, 2004.

Pensa, Patty. "No Signals, Debris Make Driving Risky." *South Florida Sun—Sentinel*, October 25, 2005.

Perrow, Charles. *Normal Accidents: Living With High-Risk Technologies*. Princeton, NJ: Princeton University Press, 1999.

Perrow, Charles. *The Next Catastrophe: Reducing Our Vulnerabilities to Natural, Industrial, and Terrorist Disasters*. Princeton, NJ: Princeton University Press, 2011.

Pestritto, Ronald J. "What America Owes to Woodrow Wilson." *Society* 43, no. 1 (2005), 57–66.

Peters, B. Guy. *Institutional Theory in Political Science: The New Institutionalism*. London: Pinter, 1999.

Petrolia, Daniel R., Craig E. Landry, and Keith H. Coble. "Risk Preferences, Risk Perceptions, and Flood Insurance." *Land Economics* 89, no. 2 (2013), 227–45.

Picou, J. Steven and Brent K. Marshall. "Katrina as Paradigm Shift: Reflections on Disaster Research in the Twenty-First Century." In *The Sociology of Katrina: Perspectives on a Modern Catastrophe*, edited by David L. Brunsma, David Overfelt, and J. Steven Picou. Lanham, MD: Rowman & Littlefield, 2007.

Pigou, Arthur C. *Wealth and Welfare*. London: Macmillan, 1912.

Pounds, Stephen and Stephanie Horvath. "Storm Weariness, Wariness Evident." *Palm Beach Post*, October 20, 2005.

Powell, Walter W. and Jeannette A Colyvas. "Microfoundations of Institutional Theory." In *The SAGE Handbook of Organizational Institutionalism*, edited by Royston Greenwood, Christine Oliver, Roy Suddaby, and Kerstin Sahlin. London: Sage, 2008.

Preuss, Lutz. "On the Contribution of Public Procurement to Entrepreneurship and Small Business Policy." *Entrepreneurship & Regional Development: An International Journal* 23, no. 9–10 (2011), 787–814.

Prier, Eric, Cliff McCue, and Ravi Behara. "The Value of Certification in Public Procurement: The Birth of a Profession?" *Journal of Public Procurement* 10, no. 4 (2010), 512–540.

Prier, Eric, Clifford P. McCue and Michael E. Bevis, "Making it Happen: Public Procurement's Role in Integrating Economic Development and Sustainability Strategies for Local Governments in the U.S.A." *3rd International Public Procurement Conference Proceedings*, 28–30 August 2008. http://goo.gl/ekfxo.

Prohaska, Sarah. "Residents Start Prepping for Wilma—Just In Case." *Palm Beach Post*, October 20, 2005.

Qiao, Yuhua, Khi V. Thai, and Glenn Cummings. "State and Local Procurement Preferences: A Survey." *Journal of Public Procurement* 9, no. 3 (2009), 371–410.

Reich, Robert B. *Aftershock: The Next Economy and America's Future*. New York: Vintage, 2011.

Reppen, Deena and Scott Openshaw. "Governor Bush Announces Emergency "Bridge Loans" Available for Small Business Severely Impacted by Wilma." November 2, 2005. http://www.dep.state.fl.us/mainpage/em/2005/wilma/news/1102_02.htm.

Reuters. "Minot Braces for Floods; A Quarter of North Dakota City Evacuates." *Calgary Herald*, June 25, 2011.

Richardson, Harry W., Peter Gordon, and James E. Moore II. *The Economic Impacts of Terrorist Attacks.* Cheltenham: Edward Elgar, 2005.

Robertson, Kent A. "Downtown Redevelopment Strategies in the United States: An End-of-the-Century Assessment." *Journal of the American Planning Association* 61, no. 4 (1995), 429–37.

Roland, Gerard. *Privatization: Successes and Failures.* New York: Columbia, 2008.

Rose, Adam. "Defining and Measuring Economic Resilience to Disasters." *Disaster Prevention and Management* 13, no. 4, 307–314.

Rosen, Harvey S. *Public Finance,* 4th ed. Chicago: Irwin, 1995.

Rubin, Claire B., Martin D. Saperstein, and Daniel G. Barbee. *Community Recovery from a Major Natural Disaster.* Boulder, CO: University of Colorado Institute of Behavioral Science, 1985.

Ruchelman, Leonard I. "Natural Hazard Mitigation and Development: An Exploration of the Roles of the Public and Private Sectors." In *Managing Disaster: Strategies and Policy Perspectives,* edited by Louise K. Comfort. Durham, NC: Duke University Press, 1988.

Runyan, Rodney C. "Small Business in the Face of Crisis: Identifying Barriers to Recovery from a Natural Disaster." *Journal of Contingencies and Crisis Management* 14, no. 1 (2006), 12–26.

Saiz, Martin. "Politics and Economic Development: Why Governments Adopts Different Strategies to Induce Economic Growth." *Policy Studies Journal* 29, no. 2 (2001), 203–214.

Sapat, Alka K. "Multiple Dimensions of Resilience: Directions for Future Research." *The Proceedings of the First International Symposium on Societal Resilience, November 30–December 2, 2010, Fairfax, Virginia.* http://homelandsecurity.org/docs/Social_Resilience_BOOK.pdf.

Schilder, Ard. *Government Failures and Institutions in Public Policy Evaluation: The Case of Dutch Technology Policy.* Assen: Van Gorcum, 2000.

Schmidt, Charles W. "Blind Rush? Shale Gas Boom Proceeds Amid Human Health Questions." *Environmental Health Perspectives* 119, no. 8 (2011), a348–a353.

Schneider, Saundra K. *Flirting with Disaster: Public Management in Crisis Situations.* Armonk, NY: M. E. Sharpe, 1995.

Schramm, Jill. "City Seeks Volunteer Planners to Lead Redevelopment Project." *Minot Daily News,* April 14, 2013.

———. "Crime, Housing, Flood Control Concerns Dominate City Council Meeting." *Minot Daily News,* August 2, 2011.

———. "Recovery Committee Reviews ESF-14 Progress." *Minot Daily News,* October 6, 2011.

———. "Slow But Steady: Valley Works toward Recovery." *Minot Daily News,* October 23, 2012.

———. "The Day the Sirens Sounded Minot Dispatch, Officials Look Back on June 22." *Minot Daily News,* January 1, 2012.

Schultz, Jason and Jennifer Sorentrue. " 'Depression-Level' Unemployment Dampens Spirits, Glades Residents Desperate for Jobs." *Palm Beach Post,* November 12, 2009.

Schumacher, Ingmar and Eric Strobl. "Economic Development and Losses Due to Natural Disasters: The Role of hazard Exposure." *Ecological Economics* 72 (2011), 97–105.

Sclar, Elliott D. *You Don't Always Get What You Pay For: The Economics of Privatization.* Ithaca, NY: Cornell University Press, 2000.

Seawright, Jason and John Gerring. "Case Selection Techniques in Case Study Research: A Menu of Qualitative and Quantitative Options." *Political Research Quarterly* 61, no. 2 (2008), 294–308.

Senge, Konstanze. "The 'New Institutionalism' in Organization Theory: Bringing Society and Culture Back In." *The American Sociologist* 44, no. 1 (2013), 76–95.

Sherif, Muzafer and Carolyn W. Sherif. *An Outline of Social Psychology.* New York: Harper & Row, 1956.

Simerman, John. "In Briana Allen Killing, Prosecutors Unleash Major Racketeering Indictment." *The Times-Picayune,* May 9, 2013.

Simon, Herbert A. "Rationality in Psychology and Economics." *The Journal of Business* 59, no. 4 (1986), S209–24.

Sims, Ronald R. *Managing Organizational Behavior.* Westport, CT: Quorum, 2002.

Slaymaker, Alex. "A New Plan B: Relocate American Cities." *The Bulletin of the Atomic Scientists (online),* June 11, 2013. http://thebulletin.org/new-plan-b-relocate-american-cities

Smith, Adam. *An Inquiry into the Nature and Causes of the Wealth of Nations.* London: Methuen, 1776/1904.

Smith, Dianna. "Affordable Housing Proposals Grab Delray." *Palm Beach Post,* May 17, 2006.

Smith, Jennifer Jo Snider. "Competition and Transparency: What Works for Public Procurement Reform." *Public Contract Law Journal* 38, no. 1 (2008), 85.

Solecki, William D. "Environmental Hazards and Interest Group Coalitions: Metropolitan Miami after Hurricane Andrew." In *Crucibles of Hazard: Mega-Cities and Disasters in Transition,* edited by James K. Mitchell. New York: United Nations University, 1999.

Southwick, Steven M. and Dennis S. Charney. *Resilience: The Science of Mastering Life's Greatest Challenges.* New York: Cambridge University Press, 2012.

Stinchcombe, Arthur L. "On the Virtues of the Old Institutionalism." *Annual Review of Sociology* 23 (1997), 1–18.

Stivers, Camilla. "So Poor and So Black: Hurricane Katrina, Public Administration, and the Issue of Race." *Public Administration Review* 67, no. 6 (2007), 48–56.

Story, Louise. "As Companies Seek Tax Deals, Governments Pay High Price." *The New York Times,* December 1, 2012. http://www.nytimes.com/2012/12/02/us/how-local-taxpayers-bankroll-corporations.html.

Suleiman, Ezra N. *Dismantling Democratic States.* Princeton: Princeton University Press, 2003.

Swan, Alonzo M. *Canton: Its Pioneers and History.* Peoria, IL: Nason, 1871. http://archive.org/details/cantonitspioneer01swan.

Taleb, Nassim. *Fooled By Randomness: The Hidden Role of Chance in the Markets and in Life.* London: Texere, 2001.

———. *The Black Swan: The Impact of the Highly Improbable.* New York: Random House, 2007.

Taylor, Shelley E. and Jonathon D. Brown. "Illusion and Well-being: A Social Psychological Perspective on Mental Health." *Psychological Bulletin* 103, no. 2 (1988), 193–210.

Thai, Khi V. "Public Procurement Re-Examined." *Journal of Public Procurement* 1, no. 1 (2001), 9–50.

Tidwell, Mike. *The Ravaging Tide: Strange Weather, Future Katrinas, and the Coming Death of America's Coastal Cities.* New York: Free Press, 2006.

Tierney, Kathleen J. "Businesses and Disasters: Vulnerability, Impacts, and Recovery." In *Handbook of Disaster Research,* edited by Havidan Rodriguez, Enrico L. Quarantelli, and Russell Dynes. New York: Springer, 2007.

Tierney, Kathleen J., Michael K. Lindell, and Ronald W. Perry. *Facing the Unexpected: Disaster Preparedness and Response in the United States.* Washington, DC: Joseph Henry, 2001.

Tran, Tu-Uyen. "Minot Changes Flood Model." *McClatchy—Tribune Business News,* February 5, 2012.

U.S. Geological Survey. "USGS Releases Updated Bakken and New Three Forks Oil and Gas Assessment." April 30, 2013. http://energy.usgs.gov/Miscellaneous/Articles/tabid/98/ID/247/USGS-Releases-Updated-Bakken-and-New-Three-Forks-Oil-and-Gas-Assessment.aspx.

U.S. National Archives and Records Administration. *Code of Federal Regulations.* Title 48. Federal Acquisition Regulations System.

Vatsa, Krishna S. "Risk, Vulnerability, and Asset-Based Approach to Disaster Risk Management." *International Journal of Sociology and Social Policy* 24, no. 10/11 (2004), 1–48.

Ward, Robbie and Emily Le Coz. "Courting the Creative Class." *Knight Ridder Tribune Business News,* October 9, 2005.

Watkins, John P. "Economic Institutions under Disaster Situations: The Case of Hurricane Katrina." *Journal of Economic Issues* 41, no. 2 (2007), 477–483.

Waugh, Richard. "Removing The Obstacles To E-Procurement Adoption." *Government Procurement* 19, no. 4 (2011), 1 Sept.

Webb, Gary R., Kathleen J. Tierney, and James M. Dahlhamer. "Predicting Long-Term Business Recovery from Disaster: A Comparison of the Loma Prieta Earthquake and Hurricane Andrew." *Environmental Hazards* 4, no. 2 (2002), 45–58.

Weber, Klaus and Mary Ann Glynn. "Making Sense with Institutions: Context, Thought and Action in Karl Weick's Theory." *Organization Studies* 27, no. 11 (2006), 1639–1660.

Weick, Karl E. "The Collapse of Sensemaking in Organizations: The Mann Gulch Disaster." *Administrative Science Quarterly* 38 (1993), 628–652.

———. *Sensemaking in Organizations.* Thousand Oaks, CA: Sage, 1995.

———. *The Social Psychology of Organizing.* New York: McGraw-Hill, 1979.

Weissmann, Jordan. "How Oil Made Working-Class North Dakota a Whole Lot Richer." *The Atlantic* (online), May 2, 2013. http://www.theatlantic.com/business/archive/2013/05/how-oil-made-working-class-north-dakota-a-whole-lot-richer/275506/.

Wetzel, Dale and Ryan J. Foley. "North Dakota City Struggles with Flood Recovery." *Charleston Daily Mail,* July 04, 2011.

Whelan, Robert K. "New Orleans: Mayoral Politics and Economic-Development Policies in the Postwar Years, 1945–86." In *The Politics of Urban Development,* edited by Clarence N. Stone and Heywood T. Sanders. Lawrence: University Press of Kansas, 1987.

Whyte, William Hollingsworth. *The Organization Man.* New York: Simon and Schuster, 1956.

Wilkie, Curtis. "Politics." In *City Adrift: New Orleans Before and After Katrina,* edited by Center for Public Integrity. Baton Rouge: Louisiana State University Press, 2007.

Wilson, James Q. *Bureaucracy: What Government Agencies Do and Why They Do It.* New York: Basic, 1989.

Wood, Debra. "Investigators Find Wilma Damage Passes Expectations." *ENR: Engineering News-Record* 255, no. 18. (2005), 14.

Woods, Michael D. and Glenn Muske. "Economic Development Via Understanding and Growing a Community's Microbusiness Segment." In *Entrepreneurship and Local Economic Development,* edited by Norman Walzer. Lanham, MD: Lexington, 2007.

Wry, Tyler Earle. "Does Business and Society Scholarship Matter to Society? Pursuing a Normative Agenda with Critical Realism and Neoinstitutional Theory." *Journal of Business Ethics* 89, no. 2 (2009), 151–171.

Xiao, Yu and Shannon Van Zandt. "Building Community Resiliency: Spatial Links between Household and Business Post-disaster Return." *Urban Studies* 49, no. 11 (2012), 2523–2542.

Zajac, Edward E. *Political Economy of Fairness*. Cambridge, MA: MIT Press, 1996.
Zolli, Andrew and Ann Marie Healy. *Resilience: Why Things Bounce Back*. New York: Free Press, 2012.
Zucker, Lynne G. "The Role of Institutionalization in Cultural Persistence." *American Sociological Review*, 42, no. 5 (1977), 726–743.
Zukin, Sharon. *Landscapes of Power: From Detroit to Disney World*. Berkeley, CA: University of California Press, 1991.
Zullo, Roland. "Public-Private Contracting and Political Reciprocity." *Political Research Quarterly*, 59, no. 2 (2006), 273–81.

# Index